Globalization and the Chinese Retailing Revolution

CHANDOS
ASIAN STUDIES SERIES:
CONTEMPORARY ISSUES AND TRENDS

Series Editor: Professor Chris Rowley,
Cass Business School, City University, UK
(email: c.rowley@city.ac.uk)

Chandos Publishing is pleased to publish this major Series of books entitled *Asian Studies: Contemporary Issues and Trends*. The Series Editor is Professor Chris Rowley, Cass Business School, City University, UK.

Asia has clearly undergone some major transformations in recent years and books in the Series examine this transformation from a number of perspectives: economic, management, social, political and cultural. We seek authors from a broad range of areas and disciplinary interests: covering, for example, business/management, political science, social science, history, sociology, gender studies, ethnography, economics and international relations, etc.

Importantly, the Series examines both current developments and possible future trends. The Series is aimed at an international market of academics and professionals working in the area. The books have been specially commissioned from leading authors. The objective is to provide the reader with an authoritative view of current thinking.

New authors: we would be delighted to hear from you if you have an idea for a book. We are interested in both shorter, practically orientated publications (45,000+ words) and longer, theoretical monographs (75,000–100,000 words). Our books can be single, joint or multi-author volumes. If you have an idea for a book, please contact the publishers or Professor Chris Rowley, the Series Editor.

Dr Glyn Jones
Chandos Publishing (Oxford) Ltd
Email: gjones@chandospublishing.com
www.chandospublishing.com

Professor Chris Rowley
Cass Business School, City University
Email: c.rowley@city.ac.uk
www.cass.city.ac.uk/faculty/c.rowley

Chandos Publishing: is a privately owned and wholly independent publisher based in Oxford, UK. The aim of Chandos Publishing is to publish books of the highest possible standard: books that are both intellectually stimulating and innovative.

We are delighted and proud to count our authors from such well known international organisations as the Asian Institute of Technology, Tsinghua University, Kookmin University, Kobe University, Kyoto Sangyo University, London School of Economics, University of Oxford, Michigan State University, Getty Research Library, University of Texas at Austin, University of South Australia, University of Newcastle, Australia, University of Melbourne, ILO, Max-Planck Institute, Duke University and the leading law firm Clifford Chance.

A key feature of Chandos Publishing's activities is the service it offers its authors and customers. Chandos Publishing recognises that its authors are at the core of its publishing ethos, and authors are treated in a friendly, efficient and timely manner. Chandos Publishing's books are marketed on an international basis, via its range of overseas agents and representatives.

Professor Chris Rowley: Dr Rowley, BA, MA (Warwick), DPhil (Nuffield College, Oxford) is Subject Group leader and the inaugural Professor of Human Resource Management at Cass Business School, City University, London, UK. He is the founding Director of the new, multi-disciplinary and internationally networked *Centre for Research on Asian Management*, Editor of the leading journal *Asia Pacific Business Review* (www.tandf.co.uk/journals/titles/13602381.asp). He is well known and highly regarded in the area, with visiting appointments at leading Asian universities and top journal Editorial Boards in the US and UK. He has given a range of talks and lectures to universities and companies internationally with research and consultancy experience with unions, business and government and his previous employment includes varied work in both the public and private sectors. Professor Rowley researches in a range of areas, including international and comparative human resource management and Asia Pacific management and business. He has been awarded grants from the British Academy, an ESRC AIM International Study Fellowship and gained a 5-year RCUK Fellowship in Asian Business and Management. He acts as a reviewer for many funding bodies, as well as for numerous journals and publishers. Professor Rowley publishes very widely, including in leading US and UK journals, with over 100 articles, 80 book chapters and other contributions and 20 edited and sole authored books.

Bulk orders: some organisations buy a number of copies of our books. If you are interested in doing this, we would be pleased to discuss a discount. Please contact Hannah Grace-Williams on email info@chandospublishing.com or telephone number +44 (0) 1865 884447.

Textbook adoptions: inspection copies are available to lecturers considering adopting a Chandos Publishing book as a textbook. Please email Hannah Grace-Williams on email info@chandospublishing.com or telephone number +44 (0) 1865 884447.

Globalization and the Chinese Retailing Revolution

Competing in the World's Largest Emerging Market

YONG ZHEN

Chandos Publishing
Oxford · England

Chandos Publishing (Oxford) Limited
Chandos House
5 & 6 Steadys Lane
Stanton Harcourt
Oxford OX29 5RL
UK
Tel: +44 (0) 1865 884447 Fax: +44 (0) 1865 884448
Email: info@chandospublishing.com
www.chandospublishing.com

First published in Great Britain in 2007

ISBN:
978 1 84334 279 3
1 84334 279 0

© Y. Zhen, 2007

British Library Cataloguing-in-Publication Data.
A catalogue record for this book is available from the British Library.

Produced from electronic copy supplied by the author.
Printed in the UK and USA.

To my parents

Contents

List of figures

List of tables

Preface

The study of retailing traditionally focuses on developed countries. The study of retailing in developing countries remains seriously underdeveloped. Chinese retailing, which services 1.3 billion people, is largely ignored. Those interested in Chinese retailing normally concentrate their attentions on individual retailers rather than systematical study. So, when I studied at the University of Cambridge in 1999, I decided to write a book about Chinese retailing. The aim of this book is to systematically understand Chinese retailing and its changes caused by the transitional economy and globalization, particularly how globalization shapes Chinese retailing; then to explore how to develop retailing in the transitional economy and how multinational retailers succeed in Chinese retailing.

Yong Zhen
Zhuhai, China
August, 2006

About the author

Kevin Y. Zhen, BA, BEng, MBA, PhD, is Assistant Professor in Business and Management at Beijing Normal University – Hong Kong Baptist University United International College (UIC) and a specialist in Chinese retailing and strategic management. He worked initially in the former Ministry of Internal Trade of China and his worked involved FDI in China. He then studied in the UK and USA, and obtained his PhD from the University of Cambridge. He is one of the first to study franchise development in China and the Chinese operations of multinational retailers, such as Wal-Mart and Carrefour. His research and consultancy is in strategic management, which he has undertaken in various industries, such as retailing, electronics, food and textiles. His papers have been published in a number of journals.

The author can be contacted via the publisher.

Acknowledgments

Many people helped make this book possible.

I feel particularly grateful to Ms. Guo Geping, the President of Chinese Chain Store and Franchise Association (CCFA); Mr. Liang Wei, the General Manager of Lianhua Supermarket Holdings Co., Ltd.; Mr. Wang Minghong, the President of the Department of International Cooperation at the Ministry of Internal Trade, P. R. China; Mr. Liu Renhao, the Vice General Secretary at SinoJapanese Food Marketing Development Committee; Professor Peter Nolan from University of Cambridge, Professor Nelson Philips from Imperial College, and many friends from WalMart, Carrefour, 7–11, and other retailers both in China, UK and USA.

I would like to thank my colleague H. P. Guo, who provided me with invaluable help in typesetting this book. I am also grateful to my family and friends whose encouragement and support made this book possible.

Yong Zhen

Acronyms

CA	Competitive Advantage
C3	Concentration Ratio by the Top 3 Companies
GATT	General Agreement on Tariffs and Trade
GATS	General Agreement on Trade in Service
CS	Convenience Store
CSA	Country Specific Advantage
DS	Department Store
EDLP	Every Day Low Price
EU	European Union
FMA	First Mover Advantage
FMCG	Fast Moving Consumer Goods
FSA	Firm Specific Advantage
GMS	General Merchandise Store
IF	The Inward Focus
JV	Joint Venture
IT	Information Technology
ICT	Information and Communication Technology
L Class	Low Income Class
M & A	Mergers and Acquisition
M Class	Middle Class
MIT	Ministry of Internal Trade
MNR	Multinational Retailer
NAFTA	North America Free Trade Agreement
NBSC	National Bureau of Statistics of China

O Class	Ordinary People Class
OF	The Outward Focus
PL	Private Label
RMB	Renminbi
SBCP	State Bureau of Commodity Price
SCM	Supply Chain Management
SCG	Single Child Generation
SM	Small and Medium sized
SMR	Small and Medium sized Retailer
SOE	State-owned Enterprise
SOR	State-owned Retailer
SPC	State Planning Commission
SMCS	Supply and Marketing Cooperative Store
WTO	The World Trade Organization

Introduction

Since China's accession to the World Trade Organization (WTO), the fact that China is becoming the "world workshop" has gradually been recognized by the world. However, becoming the world production center is just a part of China's WTO story; the other part is to ask who will control Chinese commodity circulations after the WTO, which is more important than becoming the world production center, because those who control Chinese commodity circulation, particularly Chinese retailing, will control this production center.

Since China began the economic reform in 1978, Chinese retailing has been grown at fascinating rates. Its annual growth rate reached 15 per cent. After deducting inflation issue, the annual growth rage was still over 9 per cent. In 2003, Chinese retail sales reached RMB 4.58 trillion Yuan (about US$ 554 billion), which made China the world's third largest retail market behind America and Japan. In the next decade, it is still expected to keep as high as about 10 per cent benefiting from high Chinese GDP growth, accelerating urbanization and a 1.3 billion population base.

Thus, most global retail giants regard the fast growth and the great potential of the Chinese retail market as a great opportunity; they are trying their best to explore and to take this, the world's largest emerging market. For global retail giants, the result of their competition in the Chinese market will, to a large extent, determine their positions in world retailing. Meanwhile, Chinese retailing is also critically important for China's transitional economy. This is not only because it services about one fifth of the world population, but also because those who control Chinese retailing will control this world production center and will be in a very favorable position to take full advantage of the center, such as exporting Chinese products abroad, making foreign products reach the Chinese market easily and benefiting from the fast economic growth of the world's largest emerging market. If Chinese retailing is controlled by multinational retailers (MNRs), they may not only control Chinese manufacturers on the one hand but also greatly influence the consumption of 1.3 billion Chinese on the other. The world production center without control of its own distribution channels will lose its independence in developing its national economy and be vulnerable in competing in the global market for its products.

Therefore, competing in Chinese retailing is strategically important to both foreign and Chinese companies. However, Chinese retailing had been closed for decades; since opening up, and with the flood of MNRs after the WTO, can Chinese retailing compete? There are a lot of doubts both from within and outside China. In his book *China and Global Business Revolution*, Nolan (2001) has examined different Chinese industries including aerospace, pharmaceutical, steel and coal industries; he concludes that they are very weak in their competencies and will face serious challenges after China's accession to the WTO. How about Chinese retailing? Although the market size is the third largest in the world and expected to grow by about 10 per cent each year for the next decade, 65 per cent of the national retail sales are from only 38 per cent of the national population and there are nearly 800 million Chinese whose daily retail spending is still less than US$2 per person. So, is this sector different from other sectors? Does the sector have more opportunities for the firms from developing countries to catch up than other industries? The purpose of this book is to try to answer these questions. The book is divided into four parts. The first part, comprising Chapters 1–3, examines the new trends in world retailing and discusses successful retailers' models. These provide us with a background to understand Chinese retailing. The second part, including Chapters 4–7, carefully analyzes Chinese retailing, such as the market size, the potential of the market, the structure of the industry. It concludes that before China entered the WTO, Chinese retailing was very fragmented and weak in its competitiveness. The third part, comprising Chapters 8–11, analyzes the competition between MNRs and Chinese retailers and discusses how to succeed in China. The last part, Chapters 12, discusses the impacts of the WTO on Chinese retailing, particularly the opportunities and challenges for both Chinese retailers and MNRs. Finally, the book concludes that Chinese retailing and Chinese retailers can be competitive after membership of the WTO if Chinese retailers fully take their own local advantages and right strategies, and if the Chinese government makes some fundamental changes and develops appropriate industrial policies.

Part I
The Global Retail Industry

New trends in the world retail industry

1.1 The globalization of world retailing

The last decade has witnessed many new trends emerging in world retailing, among which globalization and concentration are the most obvious. The globalization of retailing can be traced back to as early as the 1970s, when some successful national retailers, such as UK-based Marks & Spencer and Germany-based Metro, tried to extend their operations abroad. They focused mainly on the traditional Triad Market: USA, Japan and Western Europe. However, because of unfavorable conditions in the global operation at that time including higher trade barriers, expensive operation costs, little international experience, etc., many of the retailers failed, such as Marks Spencer, which failed in Canada at that time. With the accelerated globalization of the world economy and the development of technology, particularly information technology (IT), the globalization of retailing is becoming the main trend, led by Ahold and Carrefour since 1990s. From 1981 to 1990, Ahold only entered two countries; and Carrefour entered three countries. However, from 1991 to 2001, Ahold entered 23 countries and Carrefour entered 19 countries (Table 1.1). Retail giants not only enter the traditional Triad Market but also enter emerging markets, such as Mexico, China and Poland.

Table 1.1	The number of countries the main MNRs entered from 1980 to 2001			
Retailer	1981–1985	1986–1990	1991–1995	1996–2001
Ahold	1	1	3	20
Carrefour	1	3	5	14
Metro	2	3	1	9
Wal-Mart	0	0	4	5

Global operation is becoming increasingly important for a corporate survival and business growth. In 1998, among the top 200 global retailers, only 94 retailers operated internationally; while by the end of 2005, the number reached 125. In 2005, 13 of the top 20 American retailers operated internationally. The globalization trend in retailing is intensifying.

However, although retailing is going global, the globalization is just in its infancy. In 2005, there were still 75 of the top 200 global retailers remained single-country merchants; even in the USA, the world most developed retail market, only 65 per cent of the top 20 retailers operated internationally. Comparing other industries such as pharmaceutical, automotive, and electronic industries, it can be found that as early as in 2000, the top 10 largest pharmaceutical companies averagely operated in 137 countries; the top 10 largest automotive companies averagely operated in 44 countries, while the top 10 largest retailers only averagely operated in 10 countries in 2000 and 13 in 2005. So, it can be argued that there may have huge potential in retailing industry for further global operation in the next decade.

1.1.1 Why go global?

Treadgold (1990) claims that it is the perception of a relative absence of growth opportunities in the home market and the perception of identifiable growth opportunities in international markets that motivate retailers to pursue internationalization. Stonehouse et al. (2000) argue that it is the four forces: social, political, economic and technical forces that drive industrial globalization. It can be argued that the globalization of retailing is mainly because of the following reasons: First, the globalization of retailing has resulted from the globalization of the world economy, which is directly driven by the liberation of economies in former communist countries and by the deregulations in international trade and investment that were caused by the launch of the North America Free Trade Agreement (NAFTA), the creation of the World Trade Organization (WTO) from the former General Agreement on Tariffs and Trade (GATT) and the International Trade Organization (ITO) and the formation of the European Union (EU). Globalization of the world economy makes the globalization of retailing possible because it brings the globalization of production and the formation of global supply chains. The formulation of economic blocs, such as NAFTA, stimulates retailers to expand within the blocs, cross-national boundaries becoming multinational retailers (MNRs). Second, the globalization of retailing has resulted from unfavorable environments in developed countries, such as the saturation of their domestic retail markets and restrictive planning legislation, which force retailers to seek growth opportunities abroad, and the favorable conditions in developing countries, such as the deregulation of retailing policies, fragmented retail markets, rising of middle class (M Class), growing population, changing lifestyles and shopping patterns, which provide MNRs with opportunities to exploit the great market potential. The deregulation of retailing in emerging markets creates opportunity windows for global retail giants; meanwhile, the saturation of

5

retailing in developed countries motivates retailers to seek opportunities abroad to meet the request for high reward from shareholders. For example, the highly regulated environment in Western Europe forces its retailers to seek growth opportunities outside, such as in Eastern European markets. The rising M Class in emerging markets resulting from the high growth of emerging economies is formulating the segment with similar demands for similar products. Third, the globalization of retailing has resulted from the development and wide application of IT, which reduces operation costs and greatly improves management efficiency. The wide application of IT provides a powerful tool for managing the global operation with less expensive operation costs and is more rewarding, especially in supply chain management. Further, the motivations to become more competitive and more successful by achieving economies of scale and to avoid local economic downturns by being present in different markets also stimulate retailers to internationalization. Therefore, some retailers go global. Retailers' global operation not only contributes to their sales growth and economies of scale but also creates value by procurement, obtaining knowledge about new markets and new operations, optimizing cost structure and improving corporate capability. Led by the global giants Carrefour and Wal-Mart, retailing is entering the global times.

1.1.2 Where to go in the global market?

During the globalization of retailing, which market should retailers choose to enter? Wal-Mart first entered its neighbor countries, such as Mexico and Canada, followed by Argentina and Brazil in South America, then Germany and the UK in Europe, and China and Japan in Asia. Carrefour first entered Spain and other European countries, and then expanded to remote markets, such as Brazil and China. Many European retail giants often make the Eastern European countries their priority choices. From these cases, it can be found that in retail globalization, retailers normally first enter the markets that are either culturally proximal, or geographically proximal, or proximal in both; then expand to further markets in either of them (Figure 1.1.). Besides cultural and geographical issues, some other issues are also considered by the retailers in judging the attractiveness of the chosen markets. These include the regulatory environment in the markets, strategic location for further expansion, economic conditions of the markets including consumers, disposable incomes, retail sales per capita and retail profit potential in the markets, the structure of the retail markets, particularly the competitive level in the market, such as if consumer demands are met effectively. For example, many Western European retailers including Carrefour, Auchan, Metro, Ahold, Aldi, Makro, Tesco and

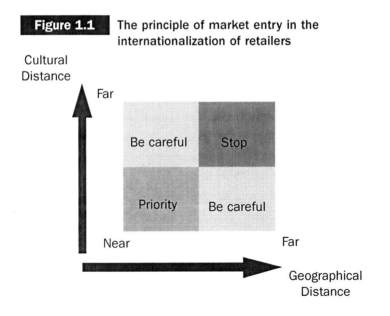

Figure 1.1 The principle of market entry in the internationalization of retailers

Rewe have entered Eastern European countries, such as the Czech Republic, Poland and Hungary. This is not only because the markets are near in both cultural and geographical distance but also because the Eastern European markets have other favorable conditions, such as the high fragment of retail markets, fast growing disposable income potential, historically less competitive environments and great consumer demands for both goods and services. In Western Europe, although the British market has lower growth potential, it is still very attractive for less intensive competition compared with other European countries, which makes the UK to have one of the highest retail profit margins in the world.

1.1.3 Obstacles and difficulties in retailers' internationalization

Over the past two decades, the global arena has proven extraordinarily difficult for many retailers going global. Few companies have succeeded in globalizing, and many barriers remain. These barriers include different institutional systems, specific government regulations, different cultures and traditions, unfamiliar rules, different distribution and logistics systems and different consumer behaviors. Further, the obstacles also include the shortage of key resources such as qualified managers; unfavourable tax and tariff structure; restriction on foreign ownership; and impenetrable established supplier relationships.

To succeed in an international market, retailers not only need to overcome these obstacles but also need to develop sustainable competitive advantages (CAs) by taking advantage of local resources. To win in international retailing, most MNRs will need to reinvent competitive advantage in each new market. Many internationalization cases fail because the retailers have failed to deal with the specific difference in economic systems in the first place, failed to understand the market difference in the second and failed to develop local CAs in the third. In addition, internationalization also involves the development of local partnerships, best practice knowledge transfer, format adaptation, export of corporate culture and back-end system integration with more centralization. Failure in any of the above issues may result in unsuccessful internationalization.

Actually, the most difficult thing for a retailer's internationalization is to understand local consumer demands and tastes, which are often influenced by the local culture. Universally appealing product assortments are always difficult to create. The difficulty in standardization of operation models makes it hard for retailers to reap economies of scale and to take central sourcing and central management in the short-term. Barth (1996) argues that the pronounced differences in consumer tastes, buying habits, and spending patterns from one country to another mean tailoring the merchandise offering along dimensions such as color, fabric, and size for apparel; brand for toys and leisure goods; and flavor for candy and snack foods. Market differences mean that a retailer profit formula can get distorted overseas; as a result, international value creation is difficult to achieve, and even more difficult to sustain. Retail business is a kind of people-intensive operation, which is hard to be managed in a different culture environment. Retailers get used to using their former success model to tailor local markets and using their successful experiences to govern their local operation. They often have the hypotheses that local consumers are similar to those in their home market, which often contributes to their unsuccessful operation. When Wal-Mart opened its first store in China in

1996, most foods displayed in the store were of western style and taste, which caused disappointing performance for the store.

1.1.4 Why does culture matter?

A main difficulty for retailers' internationalization is to understand the local market, particularly local culture, which shapes local consumer behavior. Culture could be an entry barrier or a dynamic mover in the internationalization of retailing (Herkovits 1970, Clutterbuck 1980). There are large cross-culture differences in shopping behavior, such as the Chinese using multiple senses when examining unpackaged food, and doing so far more than American and European shoppers; they also inspect more items and take more time to shop. A good understanding of culture is essential for the global success. Consumer goods, especially foods and drinks, are often culturally centered. In different cultural circumstances, one person's delicious food could be another person's poison.

Before entering China, Wal-Mart spent over 4 years studying the Chinese market by specially establishing an office in Hong Kong. When its first store was opened in China, the merchandise displayed was still less welcome by local consumers for misunderstanding local tastes, such as selling western style cooked food rather than fresh and local style food. According to the Market Development Manager of Wal-Mart China, Mr. Ma Jun, the main reason that Wal-Mart opened only 8 stores in its first 5 years in China was because it had not learned how to adapt to Chinese local markets. In retailing, culture impacts on every aspect of retail operation from product design to pricing and from distribution to promotion. Consumer taste, consumer shopping behavior and the pattern of consumption are all culture oriented. In addition, culture also plays an important role in management. For example, Chinese culture is characterized by the acceptance of hierarchy with the individual subordinated to the group. Then, horizon organization structure is normally not suitable in Chinese retailing management.

Good understanding the culture of the market entered could meet consumer demand well by providing customers with right assortments and could communicate with customers with appropriate promotion methods. Even in markets with similar culture, retailers need to pay attention to the market difference. The reason for the failures of many British retailers in the United States has been the superficial similarities of language and culture, where retailers have failed to look beyond these similarities and recognize the very different 'underlying' circumstances of competition and customer expectation (Rogers, 1991).

1.2 The concentration in world retailing

Concentration has become a global trend in retailing since the 1990s. In the USA, the most developed and the largest retail market in the world, retail concentration is very obvious. The number of retailers accounting for 20 per cent of American retail sales decreased from 33 in 1980 to 7 in 2005. In 2005, its Concentration Ratio (C3), which is defined by the percentage of the total retail sales accounted for by the top three retailers, was about 13 per cent. Wal-Mart alone took about 9 per cent of the total national market share. In Europe, the second largest retail market, the concentration is even more obvious. According to the Institute of Grocery Distribution (IGD, 2001), the top 132 retailers accounted for about 75 per cent of European retail sales in 1980, while in 1996, the number greatly decreased to 43 for the same sales. In 2000, the top 10 retailers took 37.4 per cent of the market share, and this is set to increase to 60.5 per cent in 2010, UK, Germany, France and Spain account for about 66 per cent of the total European retail sales, and in many European countries, the C3 is over 50 per cent. In Northern European counties, such as Sweden and Norway, the C3 is even over 80 per cent (IGD). In addition, the concentration is also presented by the consolidation of trading names. Carrefour converted all its European hypermarkets to its corporate name; in Mexico, Wal-Mart changed the names of acquired stores from CIFRA to its own brands.

The concentration is mainly driven by mergers and acquisitions (M&A) for two reasons: competing for market share in mature markets and being used as an entry mode strategy and/or a market development weapon in emerging markets. Since the 1990s, the maturity of retailing in most developed countries has made domestically organic growth more difficult than ever before, which forces some retailers to seek growth opportunity by M&A. In the USA, there were about 400 M&A cases involving $29 billion in 1998. In Europe, there were 90 M&A cases in 2000, 11 of which involved businesses with over €1 billion turnover. Further, M&A also steps over national boundaries: Wal-Mart took acquisitions in Canada, Germany and the UK; Ahold alone took 9 acquisitions in 7 countries in 2000. In Europe, the international M&A, which refers to at least one European party being involved and the volume is greater than $25 million, is becoming popular.

However, compared with other industries such as automotive and pharmaceutical, the concentration of retailing is still much lower. For example, the 50 largest pharmaceutical companies captured about 96 per cent of the total market share and the 50 largest automotive companies took about 92 per cent of the total market share in 2000 compared with only 20 per cent in retailing, which indicates that M&A in retailing is still at the beginning stage.

1.3 Other global retail trends

Besides the concentration and globalization, there are other main trends in retailing:

Technology, particularly Information and Communication Technology (ICT), has become more important in retailing than ever before. The development of ICT has been a very important force driving the globalization of retailing. While developments in transportation have played a major role in internationalizing industries and markets, by making it possible to transfer resources and goods between countries and continents, it is ICT, probably more than any other single factor, which has caused globalization. By examining the retail process, it can be found that ICT has become the infrastructure in retailing in recent years. Uhr and Vering (2001) argue that in the retailing world, there is the further accelerating trend in the use of standard software, and such software allows internal integration within the retail companies and the external integration with partners' Retail Information Systems (RIS) to support retailing-specific processes such as purchasing, logistics, supply, sales and production, in addition to the managerial information tailored to retailing. Supported by ICT, retailers can take Efficient Consumer Response (ECR), Supply Chain Management (SCM) and Category Management (CM); and economies of scale can be achieved with lower cost than ever before. The wide applications of Retail Management System

(RMS), Electronic Data Inter-exchange System (EDI), Electronic Point of Sale System (EPoS) and UPC barcodes have greatly reduced operation costs, improved operation efficiency and contributed to retail globalization as well as retailers' internationalization (Figure 1.2).

Increasing retail brand/Private Label (PL) goods in retailing. Retail brand or PL goods are products designed, developed and controlled by carrying the name of the retailer as their brand or by the brand owned by the retailer; the retailer contracts with a vendor to manufacture them and it promotes the brand by itself. In recent years, PL goods have developed fast and have put great pressure on proprietary brands. PL goods contribute over 30 per cent of Wal-Mart's sales and over 50 per cent of its profit; it will launch its own brand notebook in the near future. In France, over 50 per cent of Carrefour's sales carry its own brand. They form over 15 per cent of supermarket sales, and 44 per cent of grocery shoppers regularly buy PL goods (Hoch, 1996; Sethuraman, 1992). As a group, they have higher unit market share than top national brands in 77 out of 250 categories (Quelch and Hardings, 1996). Retailers like PL goods because of their potential to increase store loyalty, chain profitability, price

Figure 1.2 ICT in retail process

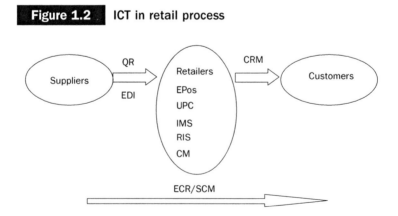

advantage (about 21 per cent cheaper than national brands), control over shelf space, and bargaining power over suppliers (Richardson, 1996).

Retail dominant power is shifting to retailers. During retailing development, the dominant power in the retail process has gradually shifted from suppliers to retailers in developed countries. Pellegrini and Reddy (1989) point out that retailers' buying power has significantly increased as a result of a process of market concentration; and wide assortments make it very difficult for manufacturers to integrate retailing, either outright or though contractual agreements, as it would imply selling products of competing firms. Hence, retailers can act independently from producers in convenience goods markets. In the current retail process, the retailer is extremely powerful and is in a favorable position in relation to suppliers and customers. This power shift is indicated by the increasing retailing concentration, the growth of PL, retailers' advantages in access to and control of the information on consumer and product sales, and the increasing control of product networks (Ogbonna and L.C., 2001). At the early stage of retailing, suppliers were more powerful than retailers and dominated retailing, because goods could not meet customer demands in quantity either due to less developed transportation or monopoly of technology by limited suppliers or less developed productivities. So retailing was less developed. With the progress of technology and its accelerating diffusion, commodities have become oversupplied, which makes retail channels more important. Thus, the dominant power in retailing gradually shifted from suppliers to retailers. The supply chain has evolved from supplier driven to retailer driven. Accordingly, customers have shifted from buying what suppliers produce to buying what retailers sell.

The increasing diversification of retail business. The diversification of retailing refers to retailers entering other retail formats or other business sectors. There is a trend of diversification in retailing in recent years. Supermarkets are offering mortgages and convenience stores are installing gas pumps. More and more retailers are becoming multi-channel operators. In 2002, the top 200 global retailers operated 2.7 retail formats compared with 1.9 in 1996. Wal-Mart not only opened hypermarkets but also warehouse and neighborhood stores. Retailers not only involve different retail formats but also develop business by other channels, such as Internet, interactive TV or direct mail. Further, driven by the increasingly intensive retail competition, many retailers start to involve other business based on their current customer bases, such as car washes, financial services, and travel service. For example, Tesco opens store restaurants; Ahold provides an insurance service; Marks & Spencer involves financial services and Carrefour develops a travel service.

The evolution of retail format is increasingly accelerated. Retail format is the type of retail mix including the nature of merchandise and service offered and its pricing policy. The development of retailing indicates the trend that the life cycle of retail format is becoming shorter and shorter. For example, it took about 100 years for department stores to mature, about 40 years for supermarkets, 20 years for hypermarkets, and 15 years for speciality stores. Now department stores are declining worldwide. Convenience stores are popular. Hypermarkets are gradually favored.

1.4 Summary

The most important development in global retailing in recent years has been retailing concentration and globalization. The maturity of retailing in developed countries has caused retailing concentration and made M&A the main path for corporate growth. Retailers are becoming larger in size while smaller in numbers. To seek to develop opportunities, some retailers take internationalization entering foreign markets, especially emerging markets. During their internationalization, cultural proximity and geographic proximity are often the key issues for choosing the market entered for low operation risks and easy operation model transfer. Meanwhile, compared with other industries, the globalization and concentration of retailing are just in their infancy, which may provide retailers from developing countries with more opportunities to catch up in this industry than other industries. In addition, there are some other significant trends in retailing such as the diversification of retailers, acceleration of retail format evolution, fast development of PL products, and so on.

The characteristics of successful retailer models

2.1 Retail strategy

Retail strategy is the overall planning guiding a retailer. It can be defined as a statement identifying: (1) the retailer's target market, (2) the retail format that the retailer plans to use to satisfy the target market's demands, and (3) the base upon which the retailer can develop its sustainable CAs (Levy Michael, 1998). To be successful, a retailer must develop an appropriate retail strategy, which means that it should identify a right target market, choose an appropriate retail format and develop its sustainable CAs; and the target market identified, the retail format chosen and the CA developed should fit each other (Figure 2.1). Because consumer demands are diverse, the retailer needs to segment its target market for focus in order to meet the demands effectively. Further, the retailer should choose a retail format fitting the target market and develop sustainable CAs recognized by the target market.

Developing sustainable CA is the core of retail strategy. Porter (1985) defines two basic types of CAs, cost advantage and differentiation advantage. The former is often associated with such retail formats as the supermarket, hypermarket, discount store and warehouse; while the latter often links with such formats as department store and specialty store. A retailer

Figure 2.1 Successful retail strategy

can achieve the cost advantage by chain operation and develop the differentiation advantages by providing customers with personalized services and leading-brand goods. However, both the cost advantage and differentiation advantage are often Inward Focus (IF), which only focuses on the value chain operation or retailer's position versus its suppliers. IF is necessary for retail success, but is not sufficient, because the CA based on IF often loses its edge with the passing of time, either because of imitation from other competitors, or because the resources for developing the CA becoming more expensive for the increasing number of companies competing for the limited resources. It is normally hard to develop sustainable CAs. Therefore, the retailer should develop a unique advantage that could be more durable while building barriers to keep out other competitors, which could be the Outward Focus (OF), focusing on developing strong consumer preference and loyalty. OF focuses on customers and direct competitors by building retail brand power.

A successful retailer should be both IF and OF, developing a strong consumer preference and loyalty based on IF while working beyond IF (Figure 2.2). IF helps a retailer become a successful retail trader; while OF makes the successful trader

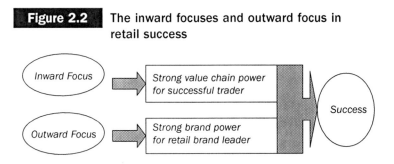

Figure 2.2 The inward focuses and outward focus in retail success

become a retail brand leader and a customer loyalty leader. A successful retailer should be a successful trader first and then become a customer loyalty leader. The former often focuses on the short-term sales and profits while the latter focuses on the long-term investment in retail brand and customer loyalty. Without strong trading ability, building strong brand loyalty is impossible; while without strong customer preference and loyalty, a retailer's success is just a momentary phenomenon and will be vulnerable in future competition. Further, a retailer without strong brand values and brand culture will often find that it is difficult to achieve internationalization.

2.2 Retailer's growth strategy

A retailer's corporate growth could be achieved either by developing economies of scale or by economies of scope or by both. Developing economies of scale requires that the retailer enters new markets in its home country or abroad, or takes market penetration to acquire more market shares from its current market. While developing economies of scope requires that the retailer diversifies its business, such as integrating its up-stream industry, developing multiple retail formats, entering a non-related business field, etc. Retailers normally make the development of economies of

Figure 2.3 Retailers' corporate growth paths

scale the main path for their corporate growth (Figure 2.3). In developed countries, retailing has matured and potential for further organic growth is very limited. Therefore, retailers often develop economies of scale for their corporate growth domestically, particularly by M&A. Besides, they can also seek growth opportunities abroad by taking internationalization. The case of Wal-Mart gives a good example for developing both successful retail strategy and corporate growth strategy; while that of Kmart provides a deep lesson in unsuccessful strategies. Chinese retailers can learn from both the success and the failure of these leading MNRs.

2.3 Case study: the success of Wal-Mart

The year 2002 saw two pieces of news that shocked the world of retailing. One was that Wal-Mart became the largest company in the world by its annual sales of US$218

Globaliza

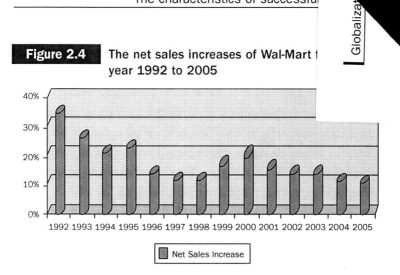

Figure 2.4 The net sales increases of Wal-Mart year 1992 to 2005

Net Sales Increase

billion. The other was that Wal-Mart's main American competitor, Kmart, which was once the third largest retailer in the US and the ninth largest in the world, filed for bankruptcy with US$4.7 billion of debt. Why do these two retail giants have such different stories? The answer may be that they have different retail strategies and different operation models.

Wal-Mart was established by Sam Walton on 31 October 1962 in Arkansas, USA. Since 1992, Wal-Mart has been maintained double-digit growth rates (Figure 2.4). In the fiscal year 2005, its total net sales reached US$312 billion, which saw Wal-Mart became the largest company in the world. The success of Wal-Mart in the USA is mainly a result of its wise retail strategy and growth strategy; and it successfully achieved both IF and OF.

2.3.1 The wise strategy: from the rural area to the urban market

When Wal-Mart started its retail business, the American retail industry had been dominated by some retail giants, such as Sears and Kmart. However, these giants concentrated in American urban areas, especially in large cities. They ignored the suburban areas and small towns, because they did not think the conditions there were suitable for opening stores unless the regional population was over 25,000. Thus, an opportunity window emerged within the industry. Sam Walton took the pre-emptive strategy of locating in American small rural towns to tap those under-exploited markets, which helped Wal-Mart not only avoid direct competition with those retail giants when it was weak but also benefit from the first mover advantage (FMA) in the markets. Sam Walton's principle was that if the population of a small town was between 5,000 and 10,000, a store could be opened there. Because the retail competition there was weak, it was easier for Wal-Mart to survive and succeed than in American large cities. Further, once a Wal-Mart store was opened there, it dominated the market, because the small town was not able to support more than one of this kind of store. At that time, surviving in the competition was critical to Wal-Mart. In American suburban areas and small towns ignored by those retail giants, Wal-Mart not only survived but also grew quickly. In 1979, its sales reached US\$1.248 billion, the first company in the USA to reach US\$1 billion in such a short time.

When Wal-Mart succeeded in the rural markets, it started to march toward American urban markets. Based on the secure base it established in its rural strongholds, Wal-Mart was able to engage in market penetration and market expansion. It took an aggressive saturation strategy to enter the urban market and expand within it, which was to fill the market entered first before developing another new market. This kind of expansion strategy was low risk and made Wal-Mart develop step by step. So, it was called the killer of American small-sized and medium-sized retailers (SMRs).

When it expanded, it always built a distribution center first, and then opened stores around the center. To meet the demands of different customer groups, Wal-Mart diversified its retail formats and developed four retail formats: Wal-Mart Discount Store, Wal-Mart Supercenter, Wal-Mart Neighborhood Market and Sam's Club; each of them targets a certain market segment and meet its demands by providing the segment with certain goods and services. By combining different retail formats, Wal-Mart filled up the market entered. By growing in the rural areas first, using the saturation strategy and duplicating the growth model for expansion, Wal-Mart became the largest retailer in the USA in 1990 and has dominated the industry since then.

2.3.2 Wal-Mart's success: a simple model without simple story

Wal-Mart's success in the USA can be summarized by its competitive model. It can be argued that Wal-Mart's success has mainly resulted from its CAs developed by the combination of its country-specific advantages (CSAs) and firm-specific advantages (FSAs) under the competition environment defined by the American government to meet its customer demands better than its competitors.

Wal-Mart's main FSAs: include its unique corporate culture, innovation capability, outstanding management know-how and strong brand image.

Unique corporate culture: Wal-Mart's unique corporate culture embodies the values of the company, which include Sam Walton's 3 Basic Beliefs, Pricing Philosophy, Sundown Rule, the Wal-Mart Cheers, Exceeding Customer Expectation, Helping People Make A Difference, etc. With these strong cultural values and norms, Wal-Mart encourages its employees to develop work behaviors focusing on providing the customer with good service. For example, the Sundown Rule states that the employees should strive to answer customers' requests by sundown on the day they receive them. The core of the corporate culture is people; just as its slogan claims "Our people make the difference." In the company, employees are called "partners" or "associates" rather than "employees." Each associate wears a badge saying: "Our people make a difference"; and besides name, there is no title for him or her, which creates an open environment free from class distinction. They are encouraged to propose suggestions and advice for improving management. Actually, the nature of the corporate culture is a kind of incentive mechanism motivating all associates to devote themselves to putting the values of the corporate culture into their daily practices. By transmitting the values to its associates by stories, myths and other means, Wal-Mart internalizes the corporate values. And the transmission is further reinforced by its organization rewards. By this organizational socialization, the associates internalize and learn the corporate cultural values so that they become organizational members. Once these values have been internalized, they become part of the individual's values and the individual follows these values even without thinking about them. The strong Wal-Mart culture motivates its associates to achieve the stringent corporate output and financial targets. So, Wal-Mart declares, "Our culture is the secret to our success and growth."

Innovation capability: Another important FSA of Wal-Mart is its innovation capability. Innovation is not only about new products but also about new ways to do business. Wal-Mart is an IT pioneer. It first implemented the barcode system in 1980 and EDI in 1985. It built the world's largest commercial electronic data system. At the end of the 1980s it was the first to use wireless scanning guns in its stores. In the 1980s it invested about US$700 million to build its information system including spending US$24 million to develop its own satellites. Wal-Mart's satellite network is the largest private satellite communication system in the world. It links all operating units of the company and its General Office with two-way voice data and one-way video communication, which makes Wal-Mart realize just-in-time (JIT) management and central management. For example, its JIT system makes the company replenish the stocks in its stores at least twice a week and many Wal-Mart stores receive daily deliveries. Its competitors, such as Kmart or Sears, only replenish their stocks every two weeks. Compared with them, Wal-Mart can maintain the same service level with about one-fourth of their inventory investment; and this is the main source of its cost saving. Wal-Mart's innovative use of IT not only contributes to its efficient operation (IF) but also allows the company to respond to different consumer demands across stores (OF). For example, Wal-Mart's materials management function is able to track the sales of individual items closely and enables Wal-Mart to optimize its product mix and pricing strategy. Then it is able to provide its customers with the right mix of goods, which not only increases the perception of value that customers associate with Wal-Mart but also makes Wal-Mart rarely left with unwanted merchandise on its hands; thus, its costs are reduced while customer loyalty is built. Therefore, by innovative use of the information system to manage its logistics, product mix inventory and product pricing, Wal-Mart develops its CA from its entire value chain and successfully achieves both IF and OF.

Management know-how: Wal-Mart's management know-how is fully embodied in its developing IF such as its cost control. The secret of Wal-Mart's success is simple, Every Day Low Price (EDLP). This is Wal-Mart's main CA. Many companies take the same cost leadership strategy, but few of them could do this simple thing as well as Wal-Mart does. Wal-Mart's low cost is from its whole value chain rather than from one or several processes of the value chain as most other retailers do. Wal-Mart's value chain includes its firm infrastructure, human resource management, technology development, inbound logistics, operations, outbound logistics, marketing and sales, and service.

Wal-Mart's value chain analysis:

- *Firm infrastructure*: It is IT that builds Wal-Mart's main infrastructure and guarantees its highly efficient operation. Wal-Mart developed the world's largest commercial data transmit system (which is even larger than that of the American National Aeronautics and Space Administration (NASA)) supported by its own satellite system, which manages its global stores, distribution centers and suppliers cooperating closely like one company and thus achieves its low cost.

- *Human resource management*: It is based on its unique corporate culture, by which the company develops its incentive system by promotion within and stock purchase plan to motivate its associates to achieve its high productivity.

- *Technology development*: Wal-Mart is a technology pioneer. It first took barcode system in 1980; first applied EDI system in 1985; first used wireless scanning guns at the end of 1980s, etc. The innovation of "cross-docking" technology contributes to its great cost reduction.

- *Inbound logistics*: In America, Wal-Mart owns a fleet of over 3,000 trucks, which distribute over 85 per cent of its merchandise through its own distribution centers. In this way, Wal-Mart not only distributes them in time but also saves 2–3 per cent transportation cost compared with its competitors.

- *Operations*: It locates stores in suburban areas and takes the saturation strategy for expansion to keep cost low. Low cost has been in each employee's mind. Its computerized inventory system gives its managers real-time information on stocks and speeds up the re-ordering of goods.

- *Marketing and sales*: It focuses on its EDLP logo with frequent promotions but does not depend on advertising much. Its advertising costs only accounts for 0.4 per cent of the total costs compared 10.6 per cent of Kmart's.

- *Service*: Wal-Mart claims to provide customers with customer-oriented service; actually it just provides customers with appropriate service to keep its cost low.

- *Strong brand imagine*: Besides IF for improving the value chain efficiency and reducing operation costs, Wal-Mart also emphasizes OF for developing its strong brand image to breed customer loyalty and preference. That "In Sam's, we trust" has become a slogan from mouths to mouths. The strong brand name is a particularly important advantage in its internationalization.

CSA based on America: Wal-Mart's strong CSAs mainly include: (1) Good economic environment. The sustainable growth of the American economy contributed to residents' income and spending and benefited Wal-Mart's business. (2) Rich human resources in management and technology. Without enough qualified managers and employees, Wal-Mart would not have been able to successfully develop 3,856 American stores by the end of the fiscal year 2006. (3) Home culture. A deep understanding of American culture allowed Wal-Mart to develop the right retail strategy. When it takes internationalization, the loss of this advantage makes it difficult for Wal-Mart to succeed in the new markets it enters. (4) Developed relative industries. The developed infrastructures in America such as the logistics industry and IT benefit Wal-Mart's success. In addition, the flexible American institutional environment, such as its planning and legislation system, which is not as strict as that in Western Europe, also contributes to its successful expansion. Thus, based on the FSAs and CSAs, Wal-Mart develops its CAs out-performing its competitors in the past 20 years (Figure 2.5).

EDLP indicates everyday low cost, which is achieved by developing economies of scale, central procurement, effective information systems and a highly efficient supply chain. Wal-Mart's chain store operation brings economies of scale. Because Wal-Mart uses central procurement, its strong bargaining power allows it to negotiate large price reductions with its suppliers, which in turn promotes its CA of EDLP. In order to keep its strong bargaining power, Wal-Mart does not depend on any vendor too much. Among

Figure 2.5   The sources of Wal-Mart's CAs

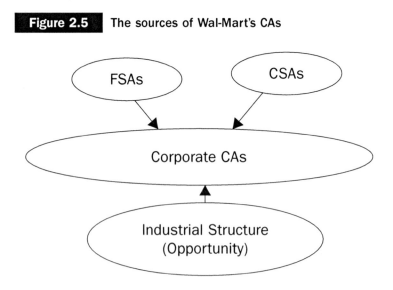

its approximately 3,000 vendors, none of them supplies over 4 per cent of the overall purchase volumes. The EDLP not only comes from its low procurement cost but also from its low cost operation in the whole operation system. Just as Lee Scott, the CEO of Wal-Mart, said, Wal-Mart's cost is low, which is not only because the goods Wal-Mart purchased are low cost; but when comparing with other retailers, the procurement cost is not much different between Wal-Mart and other retailers; Wal-Mart's low cost mainly comes from the way Wal-Mart does business, which is much lower than others. The second source of the EDLP is from Wal-Mart's central distribution supported by its developed logistics system. Wal-Mart overcomes the barrier of high transportation costs by establishing its sophisticated regional distribution centers, which can economize on inventory costs and maximize responsiveness to the needs of stores and customers. Central distribution with an advanced distribution system contributes to its low distribution costs and JIT management. Its transportation cost is just about 3 per cent of its total cost compared to about 5 per cent of its American competitors. Wal-Mart's advanced and complete logistics system, such as the cross-docking technology, contributes to its high turnover rate and makes its cost 2–3 per cent lower than the average level of the industry. 85 per cent of its commodities are distributed by its distribution centers compared with just 5 per cent for Kmart.

Another issue contributing to Wal-Mart's EDLP is its economies of scale based on sharing managerial skills across all stores. Besides organic growth, Wal-Mart also uses consolidation for corporate growth and uses a chain operation for developing cost leadership. It establishes networks linking merchandising outlets interconnected so well that they function as one large business entity. Wal-Mart's management know-how is also embodied in the way it manages people.

Wal-Mart creates a strategic control system that gives its associates at all levels continuous feedback about their performance as well as the company's performance. Its financial control system provides managers with day-to-day feedback about the performance of all aspects of the business. Through its sophisticated and company-wide satellite system, corporate managers at headquarters can evaluate the performance of each store and even each department within each store. Information about store profits and the rate of turnover of goods is provided to store managers every day; and store managers in turn communicate this information to their employees. By sharing such information, the company encourages all associates to learn the fundamentals of the retail business so that they can improve it.

Meanwhile, Wal-Mart insists on linking performance to rewards. Each manager's individual performance is reflected in pay raises and chances of promotion. Rather than hiring managers from outside, Wal-Mart routinely promotes them from within the company. Besides paying employees by their performance, Wal-Mart also uses a motivating system, such as sharing dividends with its employees, building employee funds linking their interest with the long-term development of the company and providing employees with favorable conditions to buy the company's stock. In the past 20 years its stock reward increased by 4,000 times, which gives its employees a high motivation. In addition, Wal-Mart instituted an elaborate system of controls such as rules and budget to shape employees' behavior. Each store performs the same activities in the same way, and all employees receive the same kind of training so that they know how to behave towards customers, by which Wal-Mart is able to standardize its operations, achieving major cost savings and allowing managers to make storewide changes easily when they need to do so.

2.4 The lessons from Kmart

Successful stories are often similar while stories of failure differ from one to another. It can be argued that, Kmart's failure is mainly because it could not successfully achieve its IF by developing cost advantage and OF by developing customer loyalty or differentiation advantage. In the past decade, American retailing was mainly dominated by Wal-Mart, Kmart and Target. They competed for every cent in each American city. During the competition, Target won customers by good quality with low price, and Wal-Mart built its EDLP image by its efficient cost control mechanism. However, Kmart was stuck in the middle; it lost to Target in product quality and to Wal-Mart in product price, which indicated that it was hard to build its own advantages, particularly the customer loyalty in the competition. The worst problem was that of availability; many of its products were often in short of supply. Meanwhile, it spent too much money on advertising, which made its sales greatly depend on advertisement. The recession of the American economy made Kmart's performance worse; some suppliers stopped supplying products because of payment problems. Then Kmart had to file for bankruptcy.

In detail, there are three main reasons for Kmart's failure. The first is its diversification strategy, which made Kmart lose focus and disperse its resources in competition. Since 1985, Kmart began to open speciality discount stores; but the stores could not develop advantages in both product assortment and service. Further, the service they provided could not meet consumers' expectation as to what speciality stores should do. Then Kmart lost its customer loyalty. Since 1991, Wal-Mart has out-performed Kmart in both sales and profits and became the largest discount retailer in the States. The second reason is in cost control. Unlike Wal-Mart, Kmart's distribution system was very inefficient, which caused both high cost and shortage of commodities. For example, in the third quarter of 2001, its cost reached 22.7 per cent of the sales compared with Wal-Mart's 17.3 per cent. The third is its wrong advertising strategy. Its sales greatly depended on weekly advertising. According to its CEO, Mr. Chuck Conway, this kind of advertising cost accounted for 10.6 per cent of its operation costs compared with Wal-Mart's 0.4 per cent. Heavy advertising and thousands of preprints not only cost much money but also needed enough storage for the goods. Once the availability problem emerged, the reputation of the company was damaged. When Kmart's management team felt the heavy burden of the cost, they immediately cut the cost dramatically, which caused huge customer loss. In addition, its new investment in the dot.com company, www.bluelight.com, did not achieve the expected result; the required expenditures of US$62.5 million and further US$55 million investment made the corporate financial status worse and pushed Kmart to the edge of bankruptcy. Then, Kmart was actually only a trading company just focused on short-term sales and profits. It failed to achieve its IF becoming a successful trader first; then it was impossible for Kmart to achieve the OF to build customer loyalty. Therefore, its failure is not surprising.

2.5 Summary

A retailer's successful model includes a successful retail strategy and a successful growth strategy. To develop a successful retail strategy, the retailer needs to target the right market, choose a suitable retail format, and more important, to develop its CAs based on the resources it has. Without CAs, the retailer is vulnerable in retail competition.

The right retail strategy makes the retailer achieve IF and contributes to its success. However, current success measures the results of the past performance and current profits are derived from customers won in the past. To keep its success, the retailer needs to develop OF, focusing on building long-term customer loyalty, which indicates what tomorrow's profits will be and corporate future success. To develop a successful growth strategy, the retailer needs to take a chain operation, develop multiple formats, etc.

Wal-Mart gives an excellent example in developing both successful retail strategy and growth strategy. It targets the American mass market by the format of supercenter and successfully develops the CA of EDLP achieving IF. Further, it also successfully develops the OF building customer loyalty. Meanwhile, by developing different retail formats and taking saturation strategy, it successfully achieves economies of scale in America. Comparatively, Kmart developed a wrong retail strategy, and failed to develop its CAs in the competition with Wal-Mart; its IF could not be achieved, without which OF was impossible to be developed. Therefore, the company had to face to the fate of bankruptcy.

The successful experiences of Wal-Mart and Kmart's experience of failure are very useful for Chinese retailers. Wal-Mart's saturation strategy and growth strategy as well as its EDLP are valuable for Chinese retailers in the WTO times, because the situation that Chinese retailers are facing is quite similar to that Wal-Mart faced at its early stage. Chinese retailers are much weaker than foreign global giants, and the giants all concentrate in Chinese large cities. Some Chinese retailers can take similar strategies as Wal-Mart's to grow up from Chinese rural and suburban markets to Chinese urban markets. Besides, Wal-Mart's EDLP can be effective in China since Chinese consumption level is low;

most Chinese are price sensitive consumers. Kmart's experience of failure teaches Chinese retailers the importance of developing CA and customer loyalty. Without customer loyalty, success is just a momentary phenomenon.

However, it can also be found that Wal-Mart's successful model is an American phenomenon. Its main CA of EDLP, which is based on its FSAs and CSAs, may face some challenges in its internationalization, because some of the FSAs and CSAs may not be transferred to another country. Therefore, how to develop and reshape the EDLP in international markets will be the key issue for Wal-Mart's success in its internationalization.

3

Case study: the internationalization of Wal-Mart

During the globalization of retailing, the influences of global retail giants like Wal-Mart and Carrefour are significant on emerging markets such as China. So, it is necessary to study these retail giants' behaviors, particularly their internationalization. The development of retailing globalization actually is led by these global retail giants and greatly depends on the processes of their internationalization. Among them, Wal-Mart must be studied, not only because it is the world's largest retailer but also the world's largest company by its sales. In the past decade, Wal-Mart maintained double-digit growth rates. In its fiscal year 2006, its sales reached US$312.5 billion across its 6,141 stores. It not only influences every aspect of its suppliers' operation but also takes more than a tenth of the total US imports from China. Wal-Mart is so powerful that its influences on the markets it enters can never be ignored.

3.1 What is a retailer's internationalization?

A retailer's internationalization refers to the retailer stretching its distribution system and operation out of its domestic market and doing retail business abroad, which could be a broad concept including cross-border shopping and international sourcing (Dawson, 1993) or a narrow concept referring only to market expansion excluding international sourcing (Sternquist, 1997). In this book, internationalization refers to the narrow concept.

There were several main reasons for Wal-Mart to step into the global market in the early 1990s. As a listed company, Wal-Mart faced great pressures from the stock market for both short-term and long-term returns; it also needed high corporate growth to motivate its associates, the main engine driving its CAs, because many of them had been involved in Wal-Mart's stock purchase plan. However, American retailing had reached maturity. The average retail space per capita increased from 7.5 square feet in 1983 to 19 square feet in 1993 (Miller, 1994), and intensive competition had forced many retailers out of business. The number of retailer bankruptcies increased from about 10,000 in 1983 to about 18,000 in 1992 (Swinyard, 1997). Wal-Mart had already become the national champion and its potential for further growth in America was limited, while markets outside America, such as emerging markets, provided Wal-Mart with great opportunities for international expansion, by which Wal-Mart could achieve more economies of scale. Besides, its success in America had made Wal-Mart's name a global brand, which provided Wal-Mart with the possibility to transfer its successful model and management know-how abroad. Therefore, in order to maintain its doubledigit growth, Wal-Mart decided to explore internationalization and made this its long-term strategy.

3.2 The process of Wal-Mart's internationalization

Wal-Mart began its internationalization from 1991. By 2005, it had operated in ten countries with 2,285 international stores (Table 3.1). The following section will review the main markets Wal-Mart entered by analyzing their favorable and unfavorable conditions for entry, Wal-Mart's strategies in the markets and its performances there.

3.2.1 North American market: Mexico and Canada

The old saying is America is American America. When Wal-Mart started its internationalization, it lacked experience in international operation. So, it was wise for Wal-Mart to enter a low risk market, and Mexico and Canada were the best choices. This was not only because they were near to the USA in cultural or geographical distances but also because their characteristics could benefit Wal-Mart's further expansion. Mexico was an emerging market while Canada was a mature market. The experience gained from Mexico could benefit its further accession to other emerging markets, such as South American countries;

Table 3.1 The process of Wal-Mart's internationalization

Countries entered	Time entered
Mexico	November 1991
Puerto Rico	June 1992
Canada	November 1994
Brazil	May 1995
Argentina	August 1995
China	August 1996
Germany	January 1998
Korea	July 1998
UK	July 1999
Japan	March 2002

while lessons learned from Canada could benefit its developing European markets with a similar culture, such as the UK.

Wal-Mart in Mexico. In 1991, Wal-Mart opened its first international store, Sam's Club, in Mexico City, cooperating with the largest Mexican retailer, CIFRA, and raising the curtain on its internationalization. Mexico mainly attracted Wal-Mart for several favorable reasons: (1) Great growth potential. Retail space per capita in Mexico was just 0.55 square feet in the early 1990s, compared to 19 square feet in the USA (Miller, 1994). (2) The less developed and very fragmented Mexican retailing. It was dominated by several large regional retailers and many traditional SMRs whose competencies were extremely weak when compared with Wal-Mart. Besides, Mexican retailers were not good at price competition, while Wal-Mart was. (3) Good private friendships between the two companies. Mr. Jeronimo Arango, the CEO of CIFRA, had a long and good personal relationship with Sam Walton, which provided the two companies with a solid base for their cooperation. (4) Low trade costs between America and Mexico. The NAFTA also provided Wal-Mart with good prospects for future development, such as lowering the trade barriers and tariffs between the two countries. Besides, the contiguous relationship of the two countries meant that Wal-Mart's brand was already known in Mexico; Wal-Mart already had some Mexican suppliers. However, there were some unfavorable conditions for Wal-Mart's entry. For example, Wal-Mart had no international experience and did not understand the Mexican market well; Mexico had different consumer behaviors and lower incomes than America, and the logistics system in Mexico was much less developed, etc.

Wal-Mart's strategy: Wal-Mart made a 50/50 joint venture (JV) with CIFRA, the largest retailer in the country, by which it learned operation expertise, took advantage of CIFRA's supplier relationships and tailored its own CAs. It educated CIFRA in supply chain management, especially in building distribution centers, which was critical to develop EDLP in Mexico. By 2001, it had built 10 large distribution centers with the state-of-the-art IT compared with only the 3 of Commercial Mexican, the second largest Mexican retailer.

Wal-Mart's performance: In 1997, Wal-Mart acquired the majority stakes of its JV from CIFRA. In 2000, CIFRA changed its name to Wal-Mart de Mexico de C.V. (Walmex). Walmex is the second largest listed company in Mexico and the leading retailer with about US$9.7 billion sales in 2001. Benefiting from its super-efficient distribution system, Walmex performs very well, such as in 2001, it achieved 4.74 per cent of profit margin compared with just 2.30 per cent of Commercial Mexican, one of its main competitors and the second largest retailer in Mexico (Figure 3.1).

Wal-Mart in Canada. The Canadian retail market was very different from the Mexican market because it was a developed and mature market. It had more favorable conditions than unfavorable conditions for Wal-Mart's entry. The main favorable conditions were: (1) Canadian culture was similar to American culture, so Wal-Mart needed less learning than in Mexico; (2) Wal-Mart had high brand recognition in Canada, because many Canadians lived near American borders and had already known Wal-Mart and its EDLP well; (3) although Canada was a mature market, its retailers were still very weak compared with Wal-Mart; (4) the targeted retailer, Woolco Canada, was a poor performance player and was cheap for acquisition; further, its poor performance was due to its high cost and low productivity, while Wal-Mart was good at cost control

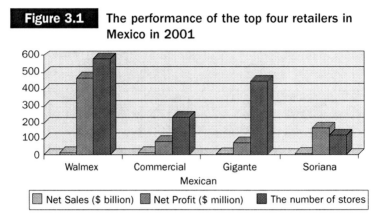

Figure 3.1 The performance of the top four retailers in Mexico in 2001

and improving productivity. More importantly, Woolco's retail format was very similar to Wal-Mart's discount store, and it had wide supplier networks and vendor-partner relationships for Wal-Mart to take. The main unfavorable conditions for Wal-Mart's entry were that new store opening was difficult in Canada because of the maturity of the industry, and the local retail competition had been very intensive.

Wal-Mart's strategy: In 1994, Wal-Mart acquired 122 of Woolco's discount department stores focusing on general merchandise. Wal-Mart sent a team to Canada for transferring Wal-Mart's way of doing business. Wal-Mart launched its "Buy Canadian" merchandising program, in which over 80 per cent of Wal-Mart's goods were sourced from local markets, which strongly contributed to its EDLP. Wal-Mart refurnished all Woolco stores up to the Wal-Mart standard.

Wal-Mart's performance: After acquired Woolco, Wal-Mart transplanted its successful model of discount stores. Just two years after the acquisition, the acquired company began to make profits. Three years later, in 1997, its market shares in the Canadian general merchandise market doubled from 22 per cent to 45 per cent. Retail sales per square foot nearly tripled to 292 Canadian dollars with greatly decreased operation costs, which attracted many Canadian retailers to hire American executives to be their CEOs or senior managers. However, because Wal-Mart just operates the format of the discount store focusing on general merchandise that has not been involved much in the food sector, its performance lags behind the top five retailers, Loblaws, Sobeys, Metro Inc, Safeway and A & P, which account for about 70 per cent of the national retail market shares.

3.2.2 South American market: Brazil, Argentina and Puerto Rico

Brazil and Argentina are the two largest retail markets in South America accounting for over 60 per cent of the retail market shares in the region. Although MNRs had entered the markets since the 1970s, the market remained fragmented. Carrefour was the largest retailer in Argentina and the second largest in Brazil. Wal-Mart first entered Puerto Rico rather than entering Brazil and Argentina. Puerto Rico was such a small market that it was ignored by many retail giants when Wal-Mart entered the market in June 1992. Wal-Mart's entry could be explained by two reasons: One was that the smaller market indicated lower risk and could be easily occupied; the other reason was that it could be used as a trial field to sharpen its international management team and to learn some new experiences of internationalization in the region, which could benefit its further expansion in the region. Because the Brazilian retail market is the largest market in the South America, this book just focuses on discussing the Brazilian market.

Wal-Mart in Brazil. After succeeding in Puerto Rico, Wal-Mart entered Brazil in May 1995. Brazil had several main favorable conditions for Wal-Mart's entry: Brazil was the largest retail market in South America with increasing purchasing power and the emergence of a middle class; the market was very fragmented and dominated by several retail giants and many small-sized retailers; there existed the long-lasting good relationships between Sam Walton and Jorge Lemann, the main shareholder and founder of Lojas Americanas, the largest retailer in Brazil.

Meanwhile, Brazil also had some unfavorable conditions for entry: (1) Wal-Mart faced strong challenges from existing large retailers in Brazil. European retailers such as Carrefour, Ahold and Makro had already built strong positions there for years. For example, Carrefour entered the market in 1975 and had developed low cost CA supported by its strong local sourcing bases and long-term supplier relationships, while thousands of small retailers had the CAs of providing good service and convenience. (2) Brazil had different culture and different consumer behaviors from America. Wal-Mart needed "tropicalization" of its product assortments. (3) There was a highly increased rate of local brand products in the market, and the oligopoly in the Brazilian manufacturing industry weakened the bargaining power of Wal-Mart. (4) In Brazil, the logistics industry was less developed. (5) Traditionally, uncertain economic performance in the region especially high inflation increased operation risk.

Wal-Mart's strategy: Wal-Mart entered Brazil by building Sam's Club JVs with Lojas Americanas, the largest discount store chain of the country under the Garantia Group. It employed Brazilian executives for the management. Wal-Mart targeted the working class and launched a differentiation strategy of low price with qualified service. It attacked Carrefour by selling comparable low price products while maintaining the advantage of better customer service, which worked to neutralize Carrefour's CAs; and attacked small retailers by the advantage of providing customers with wider merchandise assortments with comparable service, which made competitors lose their edge.

Wal-Mart's performance: Wal-Mart had a hard time for the first three years after its entry. It sold goods below their cost, which caused some suppliers to stop their supplies. Resistance to delivering goods to Wal-Mart's distribution centers caused irregular deliveries and made up to 40 per cent of products unavailable in its supercenters. Wal-Mart changed its product assortments to meet local consumer demands by emphasizing food items. Wal-Mart has built presence in the country but fails to secure significant scale. It still lags far behind Carrefour and other retailers including Pao de Acucar, CBD (under Casino), Bompreco (under Ahold), Sonae and Casas Sendas, and its financial performance remains unsatisfactory.

3.2.3 The Western European market: Germany and the UK

Before Wal-Mart entered Western Europe, the retail market there had matured. Germany, the UK and France took nearly 70 per cent of the European retail market share. The mature market was full of strong home-based competitors, such as Metro, Tesco and Carrefour, which had developed similar retail strategies to Wal-Mart's. Some European retailers, such as Ahold, Safeway and Sainsbury's, had operated in the USA for years; they knew Wal-Mart and the American retail market better than Wal-Mart knew them and the European market. Wal-Mart had weak supplier relationships in Europe. Its brand name was not as popular as in Latin America. Meanwhile, Europe had strict regulations in retailing, such as strict planning regulation and the Greenland Law, which provided Wal-Mart with limited opportunity for organic growth; further, Wal-Mart lacked experience of operating in such an environment. In addition, Europe has diversified cultures and consumer tastes. These indicated that the best way for Wal-Mart to develop its European retail market was by acquisition.

Wal-Mart in Germany. In Europe, Wal-Mart first entered Germany. Germany had the following main favorable conditions for entry: Germany was the largest European retail market accounting for about 20 per cent of European market share; it had the highest consumer spending per capita, one-third more than the European average level; modern retail formats were well developed with hypermarkets, supermarkets and discount stores; Germany was the leading player in the single currency; German consumers were very price sensitive and retail competition there mainly focused on price rather than service; German retailers provided customers with poor service, normal service in America and the UK such as free shopping bags and credit card acceptance were not popular in Germany; Germany had the strategically important geographic position: To the east, Wal-Mart could expand to the emerging Eastern European markets; while to the west, Wal-Mart could enter other Western European markets. At the same time, Germany also had some unfavorable conditions for entry: Consumers were loyal to national brand products; German retailing was mature with stagnant growth prospect, about 0.3 per cent annual growth rate in the next few years, and thin profit margins of about 1 per cent; most large retailers were family-owned or cooperatively owned, which indicated that they were difficult to acquire as they had no pressure from the stock market for short-term returns; Germany had high labor costs and strict regulations such as limitations in store opening hours, conditions for employing part-time employees, promotions and EDLP.

Wal-Mart's strategy: In January 1998, it first entered Germany by acquiring 21 Wertkauf hypermarket stores from the Mann family; the stores were similar to Wal-Mart's supercenters. Then Wal-Mart acquired 74 stores of the Spar Handles AG hypermarket chain in the same year. Wal-Mart provided German customers with cheap goods and some services; it developed the cost advantage by restructuring its supply chain system, by introducing a new scanning system in the stores and by creating a central logistics system cooperating with Tibbett and Britten. It also invested to improve its in-store appearance by refurbishing the stores.

Wal-Mart's performance: Wal-Mart has disappointing performance in Germany: The high operation costs in German stores call for more investment. While Wal-Mart's annual spending rate had been over US$200 million from 1999 to 2000 and the profit has not yet been seen. Although Wal-Mart has achieved about 2 per cent market share, German retailers still dominate the industry; the top five retailers are: Metro, Rewe, Edeka, Aldi and Tengelmann; Wal-Mart's entry caused a fierce price war, driving the thin profit margins even lower (less than 1 per cent) and making the developing of economies of scale particularly important in the competition. Wal-Mart was fined by the Germen government for its dumping of goods in September 2000. It has not adapted to the German market, the availability problem often occurred in its stores due to the resistance of its suppliers to change their deliveries from stores to distribution centers. Wal-Mart is facing hard times in Germany.

Wal-Mart in the UK. The British retail market was the second largest European retail market with the highest profit margin in the world. The market has had a trend of polarized development; competing for price at one end while competing for quality at the other. It had many favorable conditions for

Wal-Mart's entry: the UK was much nearer to the USA in culture than other European countries; it had a very developed logistics industry; the target retailer for the acquisition, Asda, which was the third largest retailer in the UK, was an ideal retailer for takeover: Asda's PL apparel was European fastest growing line of apparel with annual sales of over $830 million, which could be used for Wal-Mart's further internationalization; meanwhile, Asda was also the world largest Indian "takeaway" retailer and the UK's biggest retailer of Indian food; more important, Asda shared a similar corporate culture and operation model as Wal-Mart's supercenter. Actually Asda had copied Wal-Mart's formula for many years. Similarity is the key factor for Wal-Mart to successfully acquire Asda and achieve excellent performance. The British retail market also had some unfavorable conditions for entry, such as strict planning regulations, limited sites available for new store development, expensive land and rent costs, etc. Further, Asda traditionally focused on food items while Wal-Mart did not; Asda was just one-third of Wal-Mart's supercenter in size with a 75/25 food and non-food product mix while Wal-Mart's supercenter's mix was 40/60; the number of Asda stores could not make enough regional coverage in the market; and Asda had a high portion of PL goods, which was not Wal-Mart's strong suit. In addition, British consumers were traditionally concerned with quality and convenience rather than price.

Wal-Mart's strategy: Wal-Mart entered the UK by acquiring Asda for £6.7 million in June 1999. It re-branded Asda stores and changed its product mix to 50/50 for food/non-food. Wal-Mart focuses on non-food products operation in the market and tries to make them an important source of profit.

Performance in the UK: Wal-Mart achieves promising growth in the market and its British sales have accounted for nearly half of its total international sales. Asda has become Wal-Mart's most profitable international store.

3.2.4 Asian market: Hong Kong, Indonesia, Japan, Korea and China

Wal-Mart was involved in Asian markets as early as in 1992 when it supplied products to Japanese Ito-Yokado and Yaohan. In 1994, it cooperated with the Thailand-based company, C.P.Pokphand, and opened three JVs with Value Club membership discount stores in Hong Kong. It also entered Indonesia. However none of these ventures succeeded. It withdrew from those markets later. In 1996, it entered China by JV. In 1998, it entered Korea by acquiring four Metro stores. Its performance in China seems not optimistic.

3.3 Discussing Wal-Mart's internationalization

The internationalization process of Wal-Mart can be divided into four stages: (1) access and transfer; (2) scale and learning; (3) refinement and expertise; and (4) leverage, and can be explained by the Spiral Model (Figure 3.2).

Access and transfer
Access refers to how a retailer chooses an international market and an entry mode strategy in its internationalization; transfer refers to how the retailer transfers its CA and operation model to the market.

Access: (1) Which market is chosen to enter: From Wal-Mart's process of internationalization, it can be seen that Wal-Mart often chooses the entered market by considering the following issues:

Figure 3.2   Wal-Mart's spiral model of internationalization

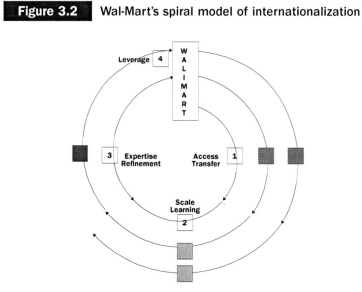

Low risk: Geographical and cultural proximity normally indicates low risk and therefore is often given priority in choosing the entry market, particularly when the retailer just starts its internationalization (Figure 3.3). Cultural proximity is important to the mass-market oriented retailer because it benefits its operation model transfer and CA transfer, while geographic proximity provides the retailer with location advantage. Wal-Mart easily stretches its distribution systems to Mexico and Canada crossing borders.

Favorable conditions for CA transfer: It is essential for the chosen market to have favorable conditions for CA transfer and development.

Great growth potential for future development: The chosen markets have at least one of the following characteristics: Fragmented local retailing such as Mexico; high profit margin with low level competition in price such as the UK; unsatisfied customer service such as Germany, which was once called the "desert of service" for its backward service.

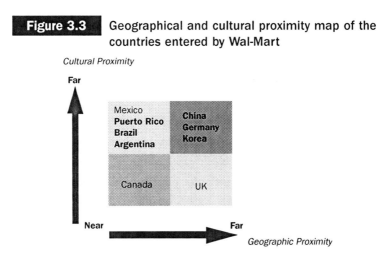

Figure 3.3 Geographical and cultural proximity map of the countries entered by Wal-Mart

Strategic position for further expansion: The chosen market has a strategic position for further development, such as China and Germany. Wal-Mart first entered Germany in Europe not only because Germany is the largest European market and the leading player in the single currency, but also because it is adjacent to central Europe and Eastern Europe, which is a strategic position for its further expansion. Wal-Mart's entry to China is not only because of China's great growth potential but also because of its rich supplies of cheap products, which is particularly important to strengthen its CA of EDLP and to support its operations both in the home market and international markets.

Besides the above issues, there is a problem of entry opportunity. For example, Canada is near America in both cultural and geographical distance, which indicates that Wal-Mart should enter the market first. The fact is that Wal-Mart did not enter Canada until 1994, three years after it entered Mexico. The main reason is that, at that time, Canadian retailing was experiencing a dramatic change period, during which many retailers were going bankrupt; Wal-Mart were waiting for the right time choosing a right target for acquisition; while Mexico was an emerging market and would take Wal-Mart a long time to adapt to it. So Wal-Mart first entered Mexico rather than Canada.

Access: (2) How to enter the chosen market: This question is about entry mode strategy. Normally there are five entry mode strategies in internationalization: JV, franchising, licencing, acquisition and building a wholly owned company. Each of them has its advantages and disadvantages. Choosing an entry mode strategy mainly depends on the retailer's business model, its strategic plan on the target market and the specific conditions of the market. In its internationalization, Wal-Mart either takes the way of new store construction (organic growth/JV) or acquisition. When it enters developed countries, such as Canada, Germany and the UK, it takes the entry mode strategy of acquisition; while entering developing countries, such as Brazil, Korea and China, it takes JV mode strategy (Figure 3.4). Compared with constructing new stores, acquisition has the advantage of taking market share faster. Acquisition could be the takeover of a strong player, such as the acquisitions Wal-Mart made in Germany and the UK, or a weak player, such as in Canada. But in both cases the target company should have a similar retail format to Wal-Mart's so that Wal-Mart can transplant its retail strategy to the acquired retailer. By acquisition or building JV, Wal-Mart transplants its corporate culture and some CAs to them. In choosing a target retailer for acquisition, retail format match is the essential condition considered. The retail format of the target retailer must match one of Wal-Mart's formats so that the transfer can be realized. Wal-Mart prefers acquisition. Its former international manager, Bob Martin said that acquisitions were a more appropriate first step in a sophisticated, well-developed retail industry. Another reason why Wal-Mart prefers expansion by acquisition is due to the pressure of short-term reward from the capital market that is floated.

Figure 3.4   Wal-Mart's entry mode strategy

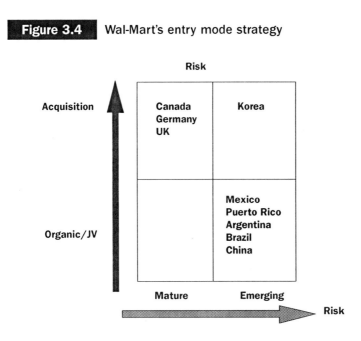

Transfer: What is transferred? How to transfer?

The essential condition for a retailer's internationalization is that it must have successful models available for transferring. Before going global, the retailer should have already developed successful business models in its home market.

(1) Retail format transfer: Wal-Mart has four retail formats: Wal-Mart Discount Store, Wal-Mart Supercenter, Wal-Mart Neighborhood Market and Sam's Club. In its internationalization, it mainly transfers Wal-Mart Discount Store and Wal-Mart Supercenter to foreign markets. In most cases, it is Wal-Mart that reshapes the acquired or cooperated retailers' formats to its formats, such as in Mexico, where it reshaped many CIFRA's former stores to Wal-Mart Discount Store and Wal-Mart Supercenter. In some cases, it transfers its retail format to the entered market as a new concept, such as transferring its Sam's Club to Mexico and China.

(2) CA transfer: In the internationalization, the transfer of CA is the most important. To transfer its CA effectively, Wal-Mart concerns "compatible" and "similarity" issues in choosing the cooperating partner for JV building and the target retailer for acquisition. The reason that Wal-Mart acquired Woolco in Canada was because it fitted Wal-Mart's discount store in retail strategy. The acquisition of value-oriented Asda in the UK was because Asda had a compatible format and similar operation model to Wal-Mart's supercenter. However, Wal-Mart's CAs are developed from the combination of its FSAs and CSAs. When Wal-Mart enters an international market, most of its CSAs are impossible to transfer because they are location-based; while some of its FSAs cannot be transferred either for the same reason or the change of consumer behavior in the new market. For example, one of Wal-Mart's CAs is its store size, which enables it to provide a wide range of product assortments to meet consumers' one-stop shopping, but when it enters Western Europe, strict planning regulation restricts the transfer of this advantage.

FSA transfer: Among Wal-Mart's FSAs, only its strong brand is non-location bound and is completely transferable. Its corporate culture, management know-how and innovation capability are in some cases location bound and only partly transferable. For example, in its corporate culture, the profit share program, which motivates its associates by providing them with Wal-Mart's shares at favorable terms, cannot be transferred in some countries because the program cannot be implemented. Its management know-how, such as marketing skills and assortment mix experiences, may not work in the new market entered for different culture and customer demands. On the technology side, the supply chain is location-bound and needs huge investment, so it is hard to transfer or duplicate in a short time. Its innovation capability is mainly based on American background, such as its deep understanding of the American market and developed American technologies, while in international markets, this capability is weakened. During Wal-Mart's inter-nationalization, the most valuable advantage or asset for the transfer actually is the brand of Wal-Mart. Bob Martin, the former international head of Wal-Mart, defines the global brand names as "low cost, best value, greatest selection of quality merchandise and highest standards of customer service."

CSA transfer: As discussed above, Wal-Mart's CSAs include a good economic environment, rich human resource in management and technology, home culture and developed relative industries. Although most of these CSAs are not transferable, Wal-Mart reduces the loss of the CSAs by choosing appropriate entry markets; for example choosing the markets with similar culture to the US market, such as Canada and UK; the markets near America in geographical distance, such as Mexico. Wal-Mart's successful model for developing CAs in the USA cannot be transferred completely to the new market entered, because local governments define a different competition environment by different regulations, and local customer demands are also different from American, and both the US-based CSAs and some FSAs cannot be transferred to the market. It must develop a new model to reach its EDLP in the international market.

Learning and scale

Learning is the process that a retailer needs in order to understand the entered market and develop new CAs by adapting to the market and taking advantage of local CSAs with its transferred CSAs. When a retailer enters a new market, it needs to consider two issues: (1) if its retail model, especially CAs, could be transferred to the new market; (2) if the CAs could be transferred, are they still CAs in the market compared with local competitors? The former issue associates with the resources supporting the CAs; while the latter involves the industrial structure in the new markets. If the home-based resources supporting its CAs cannot be transferred, or can only partly be transferred, or can be transferred but lose their edges, then the retailer must develop or reshape its new CAs. This can be done by taking advantage of local resources, which involves the learning of local markets from local competitors. From the above, it can be seen that the success of Wal-Mart is an American phenomenon; its success model cannot be duplicated completely in other countries, because of the loss of CSAs and CSAs there. Although in all countries it enters, Wal-Mart keeps its EDLP unchanged, behind the EDLP, the supply chains supporting EDLP are different, which forces Wal-Mart to learn new ways to reach its EDLP. This requires Wal-Mart to understand the local market well including local consumer demands, local suppliers, local environment, and more important, to develop effective supply relationships with local suppliers. In product mix, the standard Wal-Mart model is made up by 40 per cent hard line, 20 per cent apparel, 25 per cent dry groceries and 15 per cent fresh food. When it entered Brazil, it changed to food/non-food of 50/50 to meet local demands. In China, it learned from local retailers to provide customers with local taste foods, which are more enjoyed by them than the western-style foods Wal-Mart provided at first. When Wal-Mart enters a new market, it has to tailor its original retail format to adapt to the local market in order to overcome local resistance to its original model. For example, in Europe, after understanding the local markets, it tailored its supercenter format to hypermarket. Among Wal-Mart's formats, Supercenter and Neighborhood Market are more appropriate to the European market than Sam's Club as

European people normally do not have enough storage for bulk purchase like the Americans.

Scale: Wal-Mart develops economies of scale by increasing store opening. The fastest way to achieve this is to take acquisition and change acquired stores to Wal-Mart stores. In mature markets, it focuses on acquisition and reshaping acquired stores; while in emerging markets, it focuses on developing JV first with partners and then takes the JV from its partners. While exploring the scale, it gradually understands the market entered and keeps adjusting its operation by gradual localization. It normally opens several stores for trial and learning. When it understands the market well, it tailors its business model for further development. For example, in the first four years in China, Wal-Mart opened only eight stores for the trial. Then in the next two years, it opened eleven stores after acquiring enough knowledge about the market. The second store it opened was more attractive in product assortment to meet consumer demands better than the first one, benefiting from the learning gained from the first store. In addition, Wal-Mart can afford to lose much money when it enters a new market. It often sells goods lower than their cost to develop market share and to beat its competitors. Losses that would wipe out some competitors are mere pinpricks for Wal-Mart.

Refinement and expertise

Refinement and expertise mean Wal-Mart keeps improving its knowledge on the market entered and develops new successful retail models. By transferring its CAs to the new market, by learning new knowledge of the market entered and by adapting to local markets, Wal-Mart develops close relationships with its suppliers, customers, some competitors and non-business institutions, which contributes to the developing of its EDLP; and Wal-Mart is the center linking these relationships. Thus, the new retail model, especially new CAs, are developed.

Leverage

When Wal-Mart acquires enough knowledge and experience in a new country, it may obtain greater leverage for further expansion. In Mexico, when Wal-Mart acquired this leverage, it took the majority stakes of the CIFRA JV and then acquired the JV. When it entered Brazil, it took 60/40 JV where Wal-Mart controlled majority stakes because it had learned much international experience. The leverage helped Wal-Mart to increase internationalization from 1993 to 2002. This kind of leverage makes Wal-Mart a "glocal" retailer. Being a glocal retailer is to think globally while operate locally, which is to transfer or replicate its retail format for international expansion while adapting to local tastes with different offerings and assortments. In other words, its retail format must stay constant while the offerings and marketing methods change. Now the company has evolved into a new one with more international leverage and economies of scale, making the company more able to take further international expansion. Then, the company is able to enter other new markets from a higher level of international ability. The new round of internationalization with stages I, II and III, based on what it has learned, starts, and the internationalization is the repeating of the four stages.

3.4 Wal-Mart's international performance and problems

Since 1991, when Wal-Mart started its internationalization, its global expansion tends to be accelerated. By the fiscal year 2006, Wal-Mart owned 2,285 international stores in 10 countries and the international sales account for about 20 per cent of total sales and 25 per cent profit. Wal-Mart's internationalization has not been a smashing success. Wal-Mart has been in 10 countries and only turned profits in Mexico, Canada and the UK. In Asia, it failed and withdrew from Hong Kong and Indonesia in 1999; and its operation in Korea is worrisome. Therefore, its internationalization is still in its infancy after ten years exploration. It still lacks enough international experience and needs to continue to explore successful models in those countries. Compared with with Carrefour, the most internationalized retailer in the world, which has over 9,000 stores in 30 countries and is the leader in among 9 countries, Wal-Mart has a clear gap in financial performance.

Wal-Mart's accelerated internationalization in recent years seems increasingly makes the retail giant stuck in a dilemma. On one side, the international operation needs huge investment. Although this is not so difficult for the listed company, its annual net increase of income and current ratios for the past several years indicate that the company's financial status is not so encouraging since 1999. Its increases of net income and international sales have tended to decline, decreasing from 41.2 per cent in 2000 to 15 per cent in 2003; and its current ratio, which reflects the ability of a company to meet short-term obligations, has decreased from 1.3 in 1999 to 0.9 in 2001, which indicates that Wal-Mart may slow down its investment in international markets. Its internationalization is not optimistic.

3.5 The future of Wal-Mart's internationalization

The future of Wal-Mart's internationalization mainly depends on how Wal-Mart deals with two kinds of relationship. One is the relationship between its domestic operation and its international operation; the other, which is more important, is the relationships among different international markets. Wal-Mart's past performance shows that its home market is the most important, because about 75 per cent of its corporate profit and over 80 per cent of its sales are from America. This indicates that in future, Wal-Mart may still focus on the American market, because any problem in the market may not only influence its international operation but also the whole corporate performance. In the retail industry, there is the lesson of Marks & Spencer, which was once one of the most successful companies in the world but closed all its international stores in 2001 in order to focus on its domestic market, which was ignored during its internationalization, resulting in its poor performance.

Among Wal-Mart's international markets, the performance of each market is greatly imbalanced. Its British Asda stores contributed 44.5 per cent of Wal-Mart's total international sales and over 50 per cent of the international profit in 2002 from its 258 stores, while the other 50 per cent of sales and profit. Among its international markets, the most promising market is Mexico, where Wal-Mart has been the largest retailer, and further expansion is less risky and less costly compared with other markets. More important is that the market still has great potential for Wal-Mart's further expansion, because Mexican retailing is still very fragmented. However, Wal-Mart's operation there has some problems. Its profit margin of 4.74 per cent is even lower than the fourth largest retailer, Soriana, whose profit margin is 5.3 per cent. It is still in the regions where the former CIFRA's stores were. It has only 10 distribution centers supplying nearly 600 stores, and more distribution centers are needed for further expansion. However, it has many problems with its suppliers. The partnership relationships with them have not been built, which influences the building of distribution centers and its further expansion. Meanwhile, local retailers are learning quickly and imitating Wal-Mart. Wal-Mart still has much work to integrate and transfer the former CIFRA stores into Wal-Mart stores. In its further expansion, if Wal-Mart acquires Commercial Mexican, the second largest retailer, or Gigante, the third largest retailer, or both of them, its performance in Mexico could be improved greatly. Therefore, Wal-Mart has much work to do in Mexico.

The next promising markets are Canada, where it just has the retail format of Wal-Mart Discount Store and ranks out of the top 5 retailers, and Brazil, the largest South American market, where Wal-Mart has already built some presence. In Canada, it has not involved the food sector, which accounts for about 65 per cent of the grocery market share. If Wal-Mart introduces its supercenter to Canada, it may gain more shares from the food sector. Besides, it can be argued that the Chinese market is strategically important to Wal-Mart, because it is able to support Wal-Mart's operation in both the USA and its international markets by supplying different kinds of cheap goods and keeping its EDLP advantage; meanwhile, it is also the world largest emerging market with 1.3 billion consumers and with about 10 per cent annual retail growth rate.

In Germany, Wal-Mart's 74 stores have lost money for over 5 years. The disappointing performance there is mainly due to its small presence, less than 2 per cent of market share, the incompatibility between the corporate culture of the acquired companies and Wal-Mart's culture, and Wal-Mart's difficulty in adapting to the local environment. Wal-Mart is in a dilemma: withdraw or stay. If it stays, it has to continue to invest a lot of money to keep its operation or make another acquisition, while future performance is still uncertain. Then, the best choice may be that Wal-Mart withdraws from Germany.

International expansion requires many resources such as capital and management expertise. A company's resources are limited; while entering other markets could scatter the resources and weaken its CAs based on these resources. In China, it costs at least US$1.2 million to open one store. In Germany, US$200 million is needed each year to reshape the acquired companies. The best way to solve the capital problem in the international market is to be floated on local stock markets as Wal-Mart did in Mexico. But the case may not be repeated in each country entered. The capital problem is emerging. The internationalization is a two-edged sword; if Wal-Mart cannot deal with the two relationships, its future may be full of risks.

3.6 Summary

The process of Wal-Mart's internationalization could be divided into four stages: access and transfer, scale and learning, refinement and expertise, and leverage. During the access and transfer, Wal-Mart first entered countries with geographical and cultural proximity, and then expanded to further markets. When it enters developing countries, it normally takes JV mode; while entering developed countries it takes acquisition mode. But no matter which market it enters, its EDLP is kept unchanged. Because the conditions of the market entered are different, Wal-Mart's former successful retail model and CAs cannot be completely transferred, as it has to tailor them and develop new paths to reach its EDLP. During the scale and learning stage, Wal-Mart learns local markets further in developing economies of scale. During the third stage,

Wal-Mart realizes localization and develops expertise by formulating a new business model and CAs. During the leverage stage, the expertise obtained in developing the market is integrated into the company as a shared asset, which strengthens its ability for further internationalization.

After 10 years' exploration, Wal-Mart has achieved much progress in its internationalization, but its corporate performance still greatly depends on its domestic market that contributes over 80 per cent of the corporate sales and about 75 per cent of the corporate profit. Its international performance greatly depends on its British stores, whose profit from the 258 Asda stores is equal to the other 1,033 international stores in 8 countries. This indicates that Wal-Mart's future mainly depends on how the company deals with the relationships between its domestic market and international markets and the relationships among different international markets.

Its corporate performance indicates that Wal-Mart may still concentrate on its domestic market in the near future while using some of its international markets to support the home market and using other international markets to explore international experiences. The imbalanced performance in its international markets also indicates that Wal-Mart may focus on the promising markets, such as the UK, Mexico and Canada, to gain more profit, or focus on the bad performance markets, such as Germany and Brazil, to improve their operation. The Chinese market is important for Wal-Mart's future due to its huge supplies of cheap products and great potential and fast growing retail market. Internationalization is a two-edged sword, if Wal-Mart cannot manage it well, its internationalization might fail.

Part II
The Revolution of Chinese Retailing

4

The history of Chinese retailing

Unlike Western retailing, which evolves as a relatively independent industry under the free market economy, Chinese retailing is always under government control and administration. Most Chinese retailers have evolved either from government projects or from state-owned enterprises (SOE) operated under the planned economy. Since China began economic reform in 1980s, the Chinese retail scene has changed dramatically. The change is both revolutionary and evolutionary. In Western countries, the four retail revolutions emerged over about 100 years from 1860s to 1970s; while in China, the similar changes took place within the 10 years from 1992 to 2002. It can be argued that Chinese retailing took only about 10 years to finish the road that Western retailing took about 100 years to walk. During the emergence of new retail formats and their diffusion, the government's visible hand is always found driving the Chinese retail change. The development of Chinese retailing could be divided into three main phases by different stages of Chinese economic development, as detailed below.

4.1 Chinese retailing in the planned economy period (1949–1979)

Before 1980, the Chinese economy was a highly centralized planned economy. Because Chinese government took the heavy-industry-oriented development strategy from 1950s, its light industry had been less developed and thus resulted in great shortages of consumer goods. Since 1950s, the government had totally controlled the circulation of all goods from their production to sales. Merchandise was manufactured by store-owned factories and was then allocated by the distribution system controlled by the central government. Most Chinese earned similar low income. Each month, they were allocated a certain number of goods through the shops that they were fixed. Many daily necessities were supplied by coupons due to shortages.

During this period, Chinese retailing was completely under the government control and administration as an allocation tool. There were three government departments managing the retailing: the State Planning Commission (SPC), the Ministry of Commerce (MC) and the State Bureau of Commodity Price (SBCP). The SPC dictated overall production goals. It made production plan for each production, known as the plan quota. It also managed allocatoin of raw materials to factories according to the products they were asked to produce. After products were produced, they were submitted to the MC, which in turn were allocated to people's "work units". This process was called "fenpei" (allocation). So Chinese retailing was just a part of the whole allocation system as the extensions of the allocation channel. The SBMP was in charge of product pricing. Commercial transaction was not encouraged in the period. There was no real market and competition in retailing. MC was once one of the most powerful departments in the central government for it controlled product distribution and was the main department managing retailing. It had three departments: the Commerce Department, which managed retailing in urban areas; the Grain Department, which controlled the supplies of grains and edible oils; and the Supply and Marketing Co-operatives, which managed retailing in rural areas.

The goal of Chinese retailing was to ensure that goods could be allocated to people according to their needs and were available to those who needed them. By 1978 there were about one million retail stores as allocation outlets with 4.8 million employees serviced for nearly one billion people. Actually because of goods shortages, people's needs were never met. Each year, the SPC made plan quotas for each product. Factories then took production by the quotas. When products were produced, they were transported to the allocation system under the MC: they were firstly transported to the three national level distribution centers (in Tianjin, Shanghai and Guangzhou); then were transported to local wholesale stations; finally were transported to local stores. The distribution system was a three-tier system:

- Tier 1 – national distribution centers
- Tier 2 – provincial level wholesale stations
- Tier 3 – local stores

Products circulated along the following layouts: From Factory to National Distribution Centers to Provincial Warehouses Stations to Local Stores to Customers.

Stores sold products at the price set by the SBCP. In the whole process, it was state-owned transportation companies that took distribution works at each level. They just transported products from one place to another without involving any marketing support. During the planned economy period, there were just three main kinds of retail stores:

- *Baihuo*: The multi-story department store as both wholesaler and retailer
- *Zahuo*: Neighborhood general store
- *Zao wan men shi*: Convenience store

There was no concept of retail format. Retailers were like puppets, because they had no autonomy in the employment of their employees, the use of the profits they made, choosing suppliers and the marketing of products. They located at given sites and were supplied merchandises only from the fixed suppliers under the MC for that particular region. Retail prices were set by the SBCP. They were actually just outlets of distribution. They could not keep any profit and had to hand all profits to government.

By ownership, there were just two types of retailers: state-owned stores and collectively-owned stores. The former dominated urban areas while the later monopolized rural areas. Private-owned stores were not allowed neither were traditional free markets. Supply and marketing co-operative was the main type of collectively-owned store in China.

4.2 Chinese retailing in the planned commodity economy (1980–1991)

Since 1980, China began economic reform. The reform was first taken in the agriculture sector by developing "Household Responsibility System". The reform succeeded, which made rural people who accounted for the majority of Chinese population become richer and able to afford more expensive commodities. Then government continued to reform SOEs in cities by implementing the "Economic Responsibility System" for building incentive mechanism systems. The reform was intended to increase enterprise autonomy and to expand the role of financial incentives within the traditional economic structure, which included the introduction of profit retention, performance-related bonuses and permitted SOEs to produce outside the mandatory state plan. In February 1985, enterprises formally carried on the dual-track price policy, under which although the planned prices and planned quotas for delivery were maintained as before, enterprises were allowed to sell output in excess of their quotas with negotiated prices in markets, to plan their output accordingly after fulfilling the plan quotas, and had right to retain some profit they made. Thus the dual-track price system was established. From January 1987, the government began to promote the "Contract Responsibility System" (*chengbao zhi*). Many retailers were involved and gradually controlled by privates. Thus the rigid retailing became cracked. The economic reform taken in both rural and urban areas contributed to the great increase of residents' income, which drove the development of retailing by generating more demand for consumptions.

The development of non-SOEs and their contribution to the retailing: Many private enterprises emerged during the period, particularly in the Southern China. These non-SOEs have dominated some industries since 1990s such as in food processing and textiles industries. Some of them began to involve distribution and retail business, because it was difficult for non-state owned enterprises to enter the state-owned controlled retail channels and distribution systems. They had to develop own distribution channels to sell their products. The entry of these non-SOEs changed Chinese retailing structure not only by developing more retail channels but also by breaking down the monopoly position of SORs. They contributed to the retailing by two ways: first, they brought the SORs real competition. Since non-SOEs were out of the traditional economic structure, they had to build new circulation channels for their production and sales. Thus they were market-oriented. Second, they drove the emergence of real markets, because they bought their inputs and sold their products both by market prices; and they organized their production by demand and supply in the market. The rapid development of Chinese non-SOEs greatly contributes to the retailing by developing new retail channels and driving SORs to involve market oriented operation. From no competition to having some competition, it is a revolutionary progress in Chinese retailing.

During the period, benefing from the successful economic reform, the shortage of commodities in the past decades was resolved. Store opening became a hot business for sufficient product supply and increasing consumer demands. The saying of "open a store, you can make money quickly" reflected the increasing demand for retail outlets from both producers and customers. Retailing developed fast during the period.

There had been an explosive growth in store numbers: In 1978, there were over 1 million retail stores or stalls in China, while by 1992 the number had reached 10 million. Chinese government played a critical role in developing retailing. From 1980s to 1990s many "Vegetable Basket Projects" were launched by the government, which caused the great increases of retail stores. Meanwhile, retail employment increased from about 4 million in 1978 to over 20 million in 1987. And the total retail sales increased from RMB 136 billion Yuan in 1978 to RMB 434 billion Yuan in 1987.

Modern retail formats began to emerge in China: The first supermarket in China, Dongguan Friendship Store, was opened in March 1981 in Dongguan, a city of the Southern China, servicing for foreign customers based on the transaction of Foreign Exchange Certificate and provided imported goods and items not available in other retail outlets. Then the first supermarket chain in China, Dongguan Meijia, was born in 1990 in the same city.

The private-owned retailers (PORs) were booming: Within five years from 1980 to 1985, more than six million private shops were established. The annual growth speed was about 600 per cent. The ownership structure of Chinese retailing also changed greatly. In 1978, Chinese SOR took 92.30 per cent market share, Collectively-owned retailers took 7.50 per cent market share and private-owned retailers took only 0.20 per cent market share; while in 1987, SOR's market share decreased to 40.40 per cent, collectively-owned market share increased to 37.50 per cent and private-owned retailers' market share increased to 15.40 per cent. However, the PORs normally could not choose their locations freely. Most of them located at street corners and assigned by the local governments. The traditional free market, which was once called "the tail of capitalism" and was not allowed in the planned economy period, also became popular. By 1985, there had been about 61,000 markets in China.

Deregulation of the government control: During this period, Chinese government began to control less and less commodities. The number of its controlled commodities decreased from more than 100 categories in the early 1980s to 10 categories in the late of 1980s. The government gave retailers more freedom in pricing. For some important daily necessities, the government still set prices, which were called protected prices. For the majority of commodities, the government just set the upper price limits and the lowest price limits; thus the prices could float between them, which was called guided price. For some over-produced and non-basic consumption products, prices could be determined totally by market, which was called free market price. By 1988, the number of retail stores had reached nearly 9 million with over 20 million employees. But department stores were always crowded and customers often found it difficult to get attention from sales clerks.

4.3 Chinese retailing in the pre-WTO period (1992–2001)

During this period, the ownership structure of Chinese economy changed dramatically. Non-SOEs had achieved much progress while SOEs had been weakened greatly. The dominant status of SOEs in the national economy had been broken down. From the late 1990s, non-SOEs have contributed to over half of the national GDP.

By 1992, Chinese retailing had achieved great development. The number of retailers had reached 10 million with 24 million employees. Retailer sales reached RMB 1 trillion Yuan. By 1993, the dual-track price policy had been ended. Retailing became more market oriented than ever before. The products with planned price accounted for only 5 per cent of the total goods. However, from 1992 the macro-economic environment was deteriorating. Inflation went up dramatically due to the overheating economy and the unemployment rate increased greatly, due to SOE reform. The Chinese government had to follow a tight monetary policy and price control for some goods. Under these conditions, with the goal of creating more jobs, improving economy efficiency and modernizing retailing, China decided to open up retailing in 11 selected cities as a trial in 1992. The first step towards this the opening was tentative. In 1992, only one JV was approved, which was between Japan-based Yaohan and Shanghai Number One Department Store. The whole retail environment was still tightly regulated, for example, with price controls applied to most goods. In 1993, when the economic stituation changed for the better and price controls were largely eliminated, retail JVs became popular.

Since 1997, buyer's markets have emerged in China for oversupplies of most goods. Meanwhile, deflation came out and became the key problem of Chinese economy. According to the National Commercial Information Center, among 606 main categories of goods, 484 of them had been oversupplied. Retail price index had decreased by 3.2 per cent over several successive years; food price dropped by 4.7 per cent year to year. In 1997 and 1998, many retailers were forced out of businesses. It was the first time that bankruptcy had hit Chinese retailing, making it a historic event for the industry. The economic reform caused many workers to be laid off, leaving them in very worrying financial conditions. Future uncertainty made people reduce their consumption. Upmarket department stores then entered a difficult period. Against this background, foreign hypermarkets such as Carrefour entered China in December 1995, with Wal-Mart following in August 1996. Their chain store operation and low price strategy made them popular quickly.

A revolutionary change that happened in Chinese retailing during the period was the development of chain operation. In order to boost domestic consumption and modernize the Chinese distribution system, Chinese government decided to promote chain store operation from 1994. The Ministry of Internal Trade (MIT), which was evolved from the former Ministry of Commerce specially built a fund promoting chain operation for 14 big department stores whose annual sales were over RMB 100 million Yuan. The government also provided Chinese retailers with favorable conditions such as low-interest bank loans, tax reduction and discount rents to help them develop chain operation. Supported by the government, chain store operation was rapidly developed. In 1996, there were just about 700 chain store companies while in 1998, the figure reached 1,150, operating more than 21,000 stores. The largest national store chain was Lianhua Supermarket Ltd, which was built in 1991 and its annual sales were US$500 million in 1998 with 400 outlets and about 1,000 square meters each. In 1997, the MIT issued the "Notice on the Issuance of Opinions on the Business Scope of Chain Store," which required that chain stores must have at least 10 stores, each with the same name, selling the same sorts of goods, and centralized management and purchasing in a single distribution. It defined three forms of chain stores:

- *Direct chain operation*: one parent company owns and operates all stores.
- Voluntary chain: the stores are independent and have legal person status but managed and supplied under a contract with a chain store operator.
- *Franchise chain*: it is owned and operated by a master licensee, which uses the chain store operator's trademarks, technology and products under the license.

However, many chain stores actually just shared the same brand name while with separated management and supply systems; they had little co-ordination or consistency. Even Lianhua, the largest chain store operation in China is just a local retailer.

By 1997, China had 800 large department stores and supermarket, 1 million medium size supermarkets and department stores and 13.5 million small retailers (*China Business Daily*, 21st April 1998). On 25 June 1999, a new regulation on further opening retailing was issued, by which all provincial capitals were opened to foreign retailers. Meanwhile, the regulation asked foreign retailers operating in China to use their networks to export and promote Chinese products abroad. In December 2001, China successfully joined the WTO. Then Chinese retailing enters the WTO times.

4.4 Chinese retailing in the WTO period (2002–)

After China entered the WTO, foreign retailers' expansion in the Chinese market has been accelerated. Meanwhile, in order to defend competition from foreign entrants, Chinese retailers try their best to acquire as much market share as possible. M&A is becoming an obvious phenomenon in the industry.

4.5 Summary

Chinese retailing is always under the government control and administration. Before 1980s, Chinese retailing was just a part of the whole allocation system as the extensions of the allocation channel. There were only two types of retailers: state-owned stores and collectively-owned stores. The former dominated urban areas while the later monopolized rural areas. There was no concept of retail format and competition. Since China took economic reform in 1980s, the Chinese retailing has been changed dramatically. The change is both revolutionary and evolutionary. The Chinese retailing revolution is presented by the emergence of private-owned retailers, retail competition and new retail formats. The evolution is the diffusions of new retail formats in China. During the retail change, the Chinese government's visible hand is always found behind driving the change.

The opening-up of Chinese retailing

5.1 The opening-up of Chinese retailing

Although China officially opened its retail industry to foreign investors from 1992, some of them had already involved Chinese retail business through their mainland JVs from the late 1980s when those JVs were allowed to use local wholesalers to sell up to 30 per cent of their products made in Mainland China and to build their own distribution networks in the Mainland market. They normally sold their products to domestic wholesale distributors or sold products themselves by renting counters in local department stores, or opened specialty stores with their Chinese partners. In fact, their specialty stores also sold products made by other companies, which was illegal according to Chinese law. To open specialty stores, a foreign investor must submit a detailed business plan including a plan for foreign exchange balancing to the local governments where the stores will be opened.

In the 1980s, Chinese laws and regulations on foreign investment had a presumption that a foreign company's primary line of business was manufacturing rather than retailing. So there was no special regulation to regulate foreign investment in retailing. When a foreign company's investment in manufacturing exceeded US$30 million, it needed the approval of the State Council; if the investment was less than that figure, provincial or municipal level approval was enough. Some specialty store operators followed these regulations to do retail business in Mainland China. Because those operators were manufacturers and their retail operation was not so obvious, just as a part or an extension of their manufacturing business, the same investment approval rules were applied for both manufacturing and retailing. The companies applied the regulations on manufacturing to do retail business. Those foreign investors were mainly from Singapore, Japan and Hong Kong. Actually, Chinese encouragement of a free market in the early 1980s did not intend to permit foreign companies to enter the Chinese retail market.

In July 1992, the Chinese State Council issued a provisional national regulation on retailing drafted by the MIT exploring the experience of open Chinese retailing. The regulation chose 11 of the most developed cities as trial sites, which included Beijing, Dalian, Guangzhou, Qingdao, Shanghai and Tianjin, plus the five Special Economic Zones: Hainan, Shantou, Shenzhen, Xiamen, and Zhuhai. The regulation was very restrictive (see Appendix A): it requested that no matter how much a foreign company invested in Chinese retailing, it must obtain approval from the State Council and MNRs must take joint venture with Chinese companies. In the regulation, the Chinese government permitted that each of the 11 cities could approve no more than two retail JVs; the JVs must balance their foreign exchange through their own exports; and the imported goods of JVs were to be no more than 30 per cent of their annual sales volume. The regulation also requested that retail project proposals must receive approval from MIT and feasibility studies must receive approval from a provincial planning commission before they were implemented. The Chinese government had two goals for opening its retailing: using MNRs to promote Chinese products both in China and abroad and taking advantage of MNRs to modernize Chinese retailing and to educate Chinese retailers in improving their competency.

To support the opening of Chinese retailing, the State Bureau of Commodity Price (SBCP) issued a new "Price Management List" in 1992, in which the government reduced on a large scale and it became possible to build market prices, which provided an essential condition to open Chinese retailing. When Chinese retailing was opened, the first entrants were mainly from HK, Japan and some Asian countries. Most Western retailers just waited to see. The first retail JV approved by the State Council was Yaohan JV in Shanghai between Japan-based Yaohan Group and Shanghai No.1 Department Store, in which Yaohan International took 36 per cent equity, Yaohan Japan took 19 per cent and the Shanghai No.1 Department Store took 45 per cent; then followed Yansha Friendship Shopping City in Beijing as the second JV. In 1996, the total sales of retail JVs in China reached US$482 million, or 0.28 per cent of the national retail sales.

In 1995, the MIT issued the "Management Regulations on JV Stores," which stipulated that foreign investment in retailing must take the form of JV with Chinese companies; retail ventures could not engage in wholesale business or act as import/export agents; imported goods in any retail JV could not be over 30 per cent of its total sales; and any foreign partner was prohibited from holding more than 49 per cent of the equity in its retail JV. In 1995, the State Planning Commission (SPC) and the Ministry of Foreign Trade and Economic Cooperation (MOFTEC) jointly promulgated the "Catalogue Guiding Foreign Investment in

Retailing," in which the investment field was grouped into three categories: "prohibited," "restricted" and "encouraged". The "restricted" category was further divided into two classes: Class A and Class B. Retailing fell into the restricted category B, which indicated that no wholly foreign owned retailers were permitted, because the Chinese government thinks that retailing is very important to the Chinese economy while Chinese retailers are too weak to compete with foreign retail giants. "China is opening up, but they (Chinese government) are going to do it in a controlled way. You have to understand China. They cannot change overnight," said Charles Holley, the CFO of Wal-Mart International Division. In the new edition issued on 1st April 2002, retailing for general commodities is listed in the "encouraged" category while the retail operation of most commodities is still in the "restricted" category. The selling of some commodities, such as motor vehicles and tobacco, is still restricted. The logistics industry is listed in the "encouraged" category.

The second stage of the opening-up started from 1999. On 25th June 1999, the State Economic and Trade Commission (SETC) and MOFTEC issued the "Pilot Measures for Foreign-investment Commercial Enterprises," which established guidelines for foreign invested retail and wholesale JVs. By this regulation, all 27 provincial capitals, four municipal cities, five special economic zones and 14 cities administered independently under the State Plan, such as Qingdao and Dalian in the Mainland, were all opened to MNRs. Each city was permitted 2 or 3 large retailers. The qualifications for both Chinese and MNRs for building JVs were still strict. For MNRs, they must have over US$200 million assets, strong international sales networks and their annual turnover must be over US$2 billion in the last 3 years before applying for the JV; for Chinese partners, they must have over US$6 million assets and their sales must be at least US$36 million in the last three years before the application. At the same time, the terms of building JVs were becoming looser. A foreign retailer can take up to 65 per cent registered capital in a convenience chain store, specialty store and single brand chain store; and foreign retailer can hold the majority shares in a supermarket JV that has no more than 3 outlets. The regulation asked that the JV must use its networks to export and promote Chinese products abroad. By the pilot measures, foreign partners in small retail JVs, which are defined as having three or less chain stores, may take as much as 65 per cent stakes of the JVs in convenience store and specialty store. Foreign partners in large JVs could hold a majority stake if they obtained permission from the State Council. The condition for this was that the foreign partners should have purchased a large quantity of Chinese products and their international sales network be used to further expand the export of Chinese products. The pilot measures did not mention how many JVs are allowed to open in each city. However, it claimed that the JVs would be approved only if they comply with the commercial development plan of the city where they are located.

In December 2001, China formally entered the WTO and Chinese retailing is becoming more opened. China promised to lift nearly all limitations in retailing by December 2004. However, foreign investors are still not allowed to control those chain stores which sell products of different types and brands from multiple suppliers with more than 30 outlets according to the Report of Working Party on the Accession of China. By 2003, the number of foreign-invested retail companies in China was 350 with the total investment of US$4 billion; and 46 of them were approved by the central government.

5.2 The phenomenon of local approved JVS

During the opening-up of Chinese retailing, the Chinese central government often faces conflict with its local governments. Chinese local governments often open the industry further than the central government permits for their own interests. For example, as early as 1994, Shanghai municipal government was reported to have approved more than 300 overseas-invested retail projects including at least 40 JVs with total capital exceeding US$1 billion. According to the regulation, the State Council is the only authority office that has the right to approve this, and technically, the local governments have no right to approve foreign invested retail projects. However, cities with bureaucratic influence and self-interests such as Shanghai often over-stepped their power and had approved many JVs. By 1997, the State Council had approved only 22 foreign retail JVs while Chinese local governments had approved 277 in across nation. All of the 27 Carrefour stores were locally approved. For locally approved JVs, there were the following problems: They lacked of foreign trading rights; they could not expand outside the city where they were approved; different from nationally approved ventures, whose term was limited to 30 years, the term of locally approved JVs varied. In Shanghai, the average term for the JV was 50 years, while in Dalian, it was 17 years; they could have a wider business scope than a nationally approved JV.

When more and more "brave" foreign investors used a host of measures to set up stores where they pleased, without central government permission, the central government decided to clean up the 277 JVs and ordered that: (1) Chinese partners must hold at least 50 per cent shares in their retail JVs. In chain stores and in warehouse stores, Chinese partners must hold the majority shares of the JVs; while in Central China and Western China, the equity requirement could be as low as 40 per cent. (2) The Sino-foreign partnership could be as long as 30 years in coastal regions and 40 years in the Central and Western regions. (3) Direct selling firms must have real retail outlets. The 277 local approved foreign-invested retail JVs were reviewed in 1997 and 1998; few of them were actually shut down, because, although SETC forced those foreign investors to reduce their equity stakes in the JVs in order to accord with the pilot measures, few Chinese partners wanted to buy these stakes back with their own cash after they had sold their equity stake to their foreign partners.

5.3 Foreign chain operation in Chinese retailing

The Chinese government also put restrictions on foreign chain operations. To foreign invested chain stores, the MIT applied the existing State Council retail policy rather than making a special one. MIT has approved two JVs in chain operation as the national pilots: the Netherlands-based Makro, which is licensed to develop three warehouses with the China Native Produce Animal By-products Import and Export Corporation, and the Japan-based Ito-Yokado, which is permitted to open three retail stores with MIT's National Sugar and Wines Group Corporation. Wholly foreign-owned chain stores are prohibited. Ito-Yokado became the first foreign chain operator in China. However, foreign companies have managed to build their chain operation in different ways. For example, some foreign firms run an after-sales service operation as chains, which tend to be structured as branches of a manufacturing JV rather than as independent companies. Chain store operation makes retailers expand at low cost and low risk. Foreign clothing retailers generally take two ways to establish their chains: Licensing agreements or direct store operation. As to the licensing case, MNRs cooperate with local distribution agents or directly with Chinese retailers. Licensing agreements work like franchising operations with the brand owner receiving loyalty in return for sales training, product display, management supervision and provision of inventory. Most of the foreign fast food chains also take the form of licensing agreements. Alternatively, they can obtain direct ownership of stores by forming JVs. To supermarket chains, MIT asks that each JV supermarket must have its own retail license and may not use that of its Chinese partners'. The approval process to use a given site may require separate reviews by the local State Administration for Industry and Commerce branch (SAIC), Environment Protection Bureau, office of "Spiritual Civilization" (to approve the company's logo), the Health Bureau, etc, which often are costly and time consuming. In such a large market as China, chain operation is essential for retail success.

5.4 Summary

China officially opened its retail industry to foreign investors from 1992; and some regulations were made to manage the opening up of Chinese retailing. Chinese government intended to open up Chinese retailing in a gradual way and under its control, such as putting restrictions on foreign chain operations. However, the result is some disappointing. Most retail JVs were local approved, for example, among about 300 retail JVs, 277 JVs were local approved. Although Chinese State Council is the only authority office that has the right to approve foreign investment in Chinese retailing, Chinese local governments often over-stepped their power and approved a lot of foreign investment in the industry; they opened the industry further than the central government permitted for their own interests.

The real face of Chinese retailing

6.1 How large is the Chinese retail market?

How large is the Chinese retail market? This is a simple question, but is hard to answer. Because "the measurement of living standards in China is difficult... their needs changes and may differ from those in other cultures" (Howe, 1978). This book only discusses Chinese consumption by Chinese currency and does not involve purchasing power issues. According to the National Bureau of Statistics of China (NBSC), in 2003, the total Chinese retail sales reached RMB 4.58 trillion Yuan (about US$554 billion), the third largest retail market in the world behind America and Japan. However, the market is not as optimistic as the figure showed. It presents two pieces of obviously different pictures: the developed urban market and the less developed rural market; and there is a great structural gap between them (Figure 6.1). The Chinese urban market (including counties) accounts for over 70 per cent of the national retail sales with less than 40 per cent of the national population; while the rural market just takes about one quarter of the national retail sales with over 60 per cent of the national population. For example, in 2003, the urban market took about 76 per cent of the national retail sales with just about 39 per cent of the national population. In another words, nearly 800 million rural people accounted for only about 25 per cent of the national retail sales. The urban market size is as about 3 times large as the rural market while its population is just about 60 per cent of the rural population. So, Chinese urban residents have much higher consumption level than the rural residents; and Chinese retail sales mainly come from the minority of urban residents rather than the majority rural population in the vast rural areas. Chinese retail market is quite imbalanced.

 Figure 6.1 Chinese household consumption from 1978 to 2003

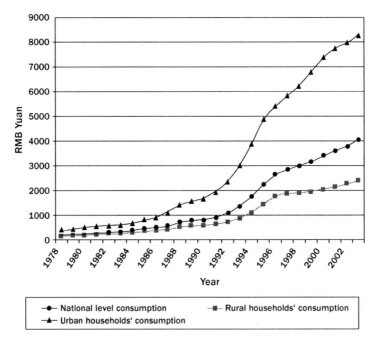

If we change a perspective to see the retail market, the imbalanced development of Chinese retailing will be more obvious. By annual retail sales per capita, the Chinese market could be divided into four sub-markets: the

Developed Market, the Relatively Developed Market, the Middle Developed Market and the Less Developed Market (Figure 6.2). The Developed Market and Relative Developed Market account for 31.2 per cent of the national retail sales by only 12.8 per cent of the national population. In Beijing and Shanghai, the retail sales per capita have reached RMB 11,530 Yuan (about US$1,390), over 3 times than the Less Developed Market including 19 provinces, which has 61.3 per cent of the national population while its sales just account for 34.4 per cent of the total. In other words, in that market with about 800 million people, each person's retail spending is less than US$2 each day. Further, although Chinese total retail sales is the third largest in the world, its retail sales per capita is extremely low compared with developed countries. In 2001, Chinese retail sales per capita was just about US$354, which was about 4.0 per cent of American, 4.1 per cent of Japanese, 6.3 per cent of British and 5.8 per cent of German sales per capita; and the figure would reach 15.6 per cent, 16.2 per cent, 24.7 per cent and 22.6 per cent separately if comparing the Developed Market with those countries.

People often say that Chinese market is large, with a market of 1.3 billion people. Is that true? From the above analysis, it can be concluded that the answer is "No". It is only a market of 495 million people including the 3 municipal cities and 9 provinces. To the Less Developed Market to the 785 million population whose retail spending per capita is less than US$2 per day, the market is not so attractive. But the 495 million people's market is attractive enough, because it has great potential for further growth; and in 2003, the areas contributed to about 70 per cent of the national GDP and its future annual growth rates are expected to be over 10 per cent in the next decade, the highest in China.

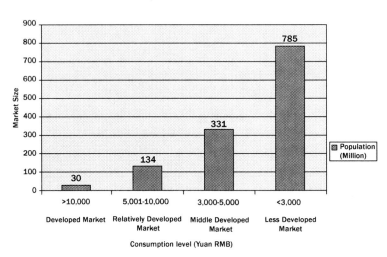

Figure 6.2 Chinese retail market size by different consumption levels

6.2 What is the potential of the Chinese retail market?

It can be argued that the future growth of Chinese retailing is mainly driven by three factors: The growth of Chinese GDP, the increase of Chinese population and the future urbanization of Chinese rural areas.

High growth rate of Chinese GDP: High and sustainable economic growth is the main force driving retailing development. Chinese sustainable growing GDP drives the increases of Chinese income, consumption and retail sales. Since China took economic reform and opening-up policy, its national economy has witnessed a dazzling growth rate. From 1989 to 2003, although the net growth of population was 165 million, the GDP per capita increased from RMB 1,512 Yuan to RMB 9,022 Yuan; the annual increase was about 8.3 per cent. The sustainable economic growth makes Chinese incomes and savings keep growing. By 2003, the total savings amount had been over RMB 10.36 trillion Yuan. Unlike Western people, Chinese get used to spend cash for shopping. Increasing incomes and saving boost Chinese spending. The national level of household consumption increased from RMB 184 Yuan in 1978 to RMB 4,058 Yuan in 2003, increased by 540 per cent by comparable price. Then Chinese retail annual growth was over 10 per cent in the past decade (Figure 6.3). The sustainable economic development for the past two decades brings the emergence and growth of a new mass market in both size and purchasing power. In the next decade, Chinese GDP is expected to keep growing at over 7 per cent each year, which will still be the main engine driving its retail growth.

Increase of population: Although Chinese population growth has been slowed down due to the "One Child Policy", its net increased population each year is still about 10 million. The total population is expected to reach 1.4 billion by 2010. In 2001, Chinese retail per capita was US$354. Even if this figure keeps unchanged, the annual

| **Figure 6.3** | The increasing gap between national savings and retail sales from 1978 to 2003 |

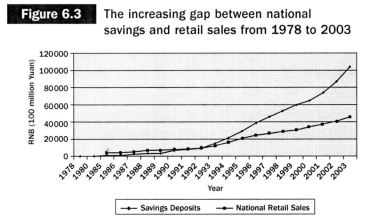

increase of retail sales from the increasing population would be US$3.54 billion. The increasing population will provide great potential for Chinese retailing.

Increasing urbanization: By the end of 2003, China had a population of about 1.3 billion, among which nearly 800 million were rural residents. According to Mr. Qiu Xiaohua, the President of the NBSC, the rural residents income per capita was RMB 2,366 Yuan in 2002 while the urban residents' was RMB 6,860 Yuan; because 40 per cent of the rural incomes were goods and 20 per cent of them would be used for reproduction, plus kinds of welfares of urban residents, the real income gap between Chinese urban and rural residents was at least 5:1. If the income gap reduces to 2:1 by the urbanization and the urbanization rate increases to 50 per cent by 2010 as per government targets, then the increase of the retail size will be huge, at least RMB 100 billion Yuan. As the urbanization accelerates, China's 800 million rural residents will be a powerful force driving Chinese retailing development and will create a fantastic opportunity for retailers. According to Mr. Huang Hai, the President of the former Internal Trade Bureau under the SETC, which was in charge Chinese retailing, the annual growth rate of Chinese retail market could be 10 per cent in the next decade. By this rate, Chinese retail sales per capita will reach US$557 in next five years and US$905 in next ten years. Considering the net population growth, the net total increase of Chinese retail sales will reach at least US$365 billion in five years and US$873 billion in ten years accordingly, which equals about 1.03 times and 2.5 times of British total retail sales in 2001 separately. If consider the urbanization issue, the increase will be more. Therefore, if the Chinese economy keeps the current growth rate, the net increases of Chinese retail market size in next decade will be huge.

6.3 The structure of Chinese retailing during economic reform

6.3.1 The dual-structure of Chinese retailing

Chinese retailing presents a typical dual-structure: The developed urban retail market co-exists with the less developed rural retail market; and the modern retail format coexists with the traditional open market. In the urban markets, retailing is more developed; diversified retail formats, such as the department store, specialty store, hypermarket, shopping mall, warehouse and convenience store share the market together. While in the rural market and small cities, it is thousands of traditional markets that dominate; there are many scattered, family-owned, SM independent retailers there. In 2003, retail sales in the urban areas accounted for about 76 per cent of the national retail sales while the rural areas took about 24 per cent (Figure 6.4).

Figure 6.4 The Chinese market shares of urban market and rural market

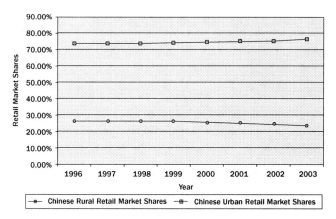

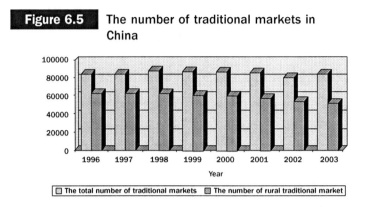

Figure 6.5 The number of traditional markets in China

There were over 81,000 traditional markets in China. This dual structure is also reflected in the geographic distribution of retailers. Eastern and Southern China has the most developed retail markets while Western China is the least developed. In 2003, the 12 Western provinces, with about 72 per cent of the national area and 30 per cent of the national population, just accounted for about 18 per cent of the national retail sales. Modern retail formats concentrate in Eastern and Southern China, and about 80 per cent of MNRs are also found there.

6.3.2 The fragmented structure

Chinese retailing is very fragmented. In 2003, the C3 was only 1.8 per cent and C10 was 3.6 per cent; even the top 100 retailers only took 9 per cent of the national sales. The largest Chinese retailer, Brilliance Group, which was merged by four large companies in April 2004, contributed to RMB 48.5 billion Yuan, 1.8 per cent of the national sales. The fragmentation also is indicated by the number of retailers per million residents. In the USA, for every one million residents there are about 2,500 retailers; while the figure reaches nearly 16,000 in China, which indicates that

Chinese retailing is just at the infant stage and consolidation will take place in future.

6.3.3 The structure of Chinese retail formats

All modern retail formats have emerged in Chinese retailing.

Department Store (DS). Since 1949, it is the independently operated department store that dominates Chinese retailing. This has been the best-established and the most popular retail format in China and still dominates Chinese retailing. In holidays it is still the most popular place for shopping. The industry is made up of thousands of this kind of DS. DS has won consumer loyalty and been their one-stop retailer for decades. It ever achieved the highest rate of space productivity in the world. Wangfujing Department Store in Beijing achieved sales of US$11,500 per square meter comparing a weight average of US$3,600 for British stores. However, after 1992, foreign entrants broke the structure when China opened its retailing. New retail formats, such as supermarket, hypermarket and warehouse, are gradually being introduced to China by MNRs. During 1990s, nearly all modern retail formats had emerged in China. Although DS still dominates Chinese retailing, its market share is shrinking. Most of them are massive single stores and not operated as chain stores. There are about 1,000 large DSs, whose area is over 5,000 square meters, and about one million SM DSs. In 2003, the top 100 DS accounted for 2.3 per cent of the national retail sales. The largest operator was Dalian Commercial City Group Corporation, whose sales reached RMB 18 billion Yuan in 2003 ranking the second in Chinese retailers. The development of the DS can be divided into four stages:

1. *The slow development stage (before 1990)*: During this period, China was still in the shortage of commodity times

and Chinese income level was very low. Retailing was just at its beginning stage. In that seller's market, little competition was involved and product promotion was not necessary. The number of large DSs was less than 100. DS developed very slowly.

2. *The fast development stage (1990–1995)*: This period was DS's golden times in China. Both commodities and people's incomes were surging. Nearly all stores did very good business. DS increased greatly; the number of large stores with annual sales over 100 million, increased from 94 to 624 . Their decorations were becoming more and more luxurious. Just in Beijing, there were over 20 large DSs whose area was over 10,000 square meters. The average net profit margin of the top 100 large DSs reached 10.10 per cent in 1994, the highest in history.

3.*The mature stage (1996–2001)*: During this period, the Chinese retail market had become a buyer's market while DS had been oversupplied. In 1997, the format reached its peak. The number of DSs whose annual sales were over RMB 120 million Yuan reached 1,000. The homogeneity of DSs caused frequent price wars, which was the only way for competition; its profit margin decreased from 10.10 per cent in 1994 to 2.74 per cent in 1997. In 1996, for the first time in China, large DSs closed their doors including some foreign invested DSs. 1998 was called "the year of closing door for department stores." Some DSs had to change to other retail formats such as supermarkets or specialty stores.

4. *The post department store times (2002–)*: Since 2002, the Chinese DS has been in decline. With China entering in the WTO, it is facing both intra-type and intertype competition from Chinese retailers and MNRs. Department store was the one-stop shopping place; while it is facing competition from specialty store for upmarket and competition from both hypermarket and supermarket for

the mass market; its market share is shrinking further. It needs repositioning. How to survive and revitalize its business are the practical problems that each Chinese department store has to face.

In China, the more developed a region is, the less the market share DS has there; and vice versa. According to the National Commercial Information Center (NCIC), the average profit margin of the DS in Shanghai, the most developed city in China, was just 2.7 per cent in 2000; and its market share was just 23 per cent, the lowest in its history. In 2003, the average profit of large department store whose retail sales was over RMB 100 million Yuan was RMB 20 million Yuan. The Chinese DS is facing many serious problems:

1. *Although since 1994, many DSs claim to have introduced chain operation, few of them have actually done so.* Their operation still remains independent, without central management.

2. *Most DSs are state-owned or controlled by the government.* The same as other state-owned enterprises (SOEs), these retailers have the problems of rigid management systems, aged employees and a heavy burden for retired employees. For example, my research finds that the average age of employees in DS is about 13 years older than that in other retail formats and the age of its management team is about 11 years older.

3. *Few of them use advanced IT in their operation management, such as in ordering, payment, inventory management and distribution.* Its core competency is weak.

4. *Many of them are stuck in their corporate culture.* They normally have very good brand and reputation as a result of the past decades' operation. But how to develop their corporate culture and adapt branding to the new environment, such as new consumer demands and new economic situations, is still unclear. They are losing their direction in intensive competition.

Supermarket. The first supermarket in China, Dongguan Friendship Store, was opened in March 1981 in Dongguan, Guangdong Province. The first supermarket chain, Dongguan Meijia, emerged in 1990 in Dongguan, too. In 1991, Lianhua set up the first supermarket in Shanghai. Since then, the format has become popular and spread throughout the whole country quickly. The supermarket chain has become a main player in Chinese retailing. In October 2000, Shanghai Hualian became the first supermarket operator that was listed on the Chinese stock market. Because the supermarket is a new format in China, it uses many new technologies and modern management experience learned from MNRs. While competing with foreign retail giants such as Wal-Mart and Carrefour, Chinese supermarkets grew quickly and developed their CAs gradually. However, during its early development stage, high price, lack of scales and small customer segment resulted in its failure in the 1980s. Most of them closed for business, with only one supermarket surviving in Shanghai and 33 in Beijing during the period. Therefore, Blois (1989) pointed out that the supermarket was not appropriate to the Chinese market because of Chinese lifestyle and Chinese economic conditions. However, the supermarket has developed quickly since the 1990s. From 1997 to 1999, its numbers tripled while DS only increased by 13 per cent. In 1994, its market share was just about 1 per cent; now it takes 6 per cent of national retail sales. It is estimated that by 2005, the supermarket will take over 30 per cent market share and replace DS to become the main retail format in China. Since 1999, Shanghai Lianhua Supermarket has become the largest retailer and replaced Shanghai No.1 Department Store. Its sales in 2003 reached RMB 24 billion Yuan. But there are some obstacles preventing Chinese supermarket's chain operation:

1. Qualified store management personnel are a scarce resource. Retailers often have difficulty finding enough management personnel for their chain operation; while training them up from the inside company takes a long time.

2. China's logistics industry is much undeveloped. There are few third party logistics companies operating nationally in China. Building their own distribution system is so expensive that most retailers are not able to afford it by themselves.

3. Local protectionism makes chain operation across regions quite difficult.

4. Most supermarkets do not have successful models for their chain operation and face difficulties in simplification, standardization and specialization.

Convenience Store (CS). The first Chinese convenience stores emerged in 1993, promoted by local governments as the governments' project. But the project soon failed because of poor management. The current convenience stores were mainly developed later. They are concentrated in the most developed cities, such as Shanghai, Guangzhou and Shenzhen. In the inland cities, convenience stores did not emerge until late 1998. The number of stores is very small and most of them are private-owned.

Shanghai is the most developed city for Chinese CS. In China, all large CS operators, whose store number is over 500, locate in Shanghai. In 2003, the total number of CS in Shanghai was 4,866 with the total sales of RMB 6 billion Yuan; Lianhua, also the largest convenience store operator, owned 1,390 CV stores.

The main foreign CS operators include Japanese Lawson and 7–11, and American Am/Pm. 7–11 entered China in 1992 by Hong Kong Dairy Farm, which has obtained the right to open convenience stores in Southern China. In Beijing, a JV has been launched by Japanese Ito-Yokado, Taiwan based President Group and a Chinese local retailer for opening 7–11 store. American "OK" plans to open 100 stores within 3 years by 2005. CS is one of the most promising formats for the future. The successful story of 7–11 in Japan may be repeated in China. The CS is normally a franchising operation, but in China, most of them are directly invested. It is just in its introduction period and only concentrated in Eastern and Southern China, the most developed regions of the country.

Hypermarket. A significant phenomenon in Chinese retailing in recent years is the development of the hypermarket. It is becoming the fastest developing, most popular and largest retail format in Chinese large cities. Its size normally is between 7,000 and 12,000 square meters. This value-oriented format is favored and booming rapidly in China. MNRs, such as Wal-Mart and Carrefour, introduced this format to China. Its main CAs are low price and wide product assortments. In fast moving consumer goods (FMCG), the hypermarket has absolute advantages over other retail formats. It has become the main competitor of all Chinese retail formats, particularly the traditional market. MNRs dominate this format and develop ahead of Chinese retailers. The main operators include Lianhua, Wal-Mart, Carrefour, Auchan, Lotus, Hy-mart, RT-mart and Trust-mart. By 2003, Carrefour had opened 41 hypermarkets in China and become the largest hypermarket operator; Wal-Mart operated 33 supercenters.

Warehouse. These first emerged in China in 1993 and there were been about twenty warehouse operators by 2002. However, most of them just have two or three stores. This format is only at the beginning stage. Its performance is not satisfactory due to positioning reasons, because it targets the same customer segment as the hypermarket, and competition from the hypermarket is strong. Wal-Mart's Sam's Club and German Metro are the main players. Metro is the largest operator; it had 18 warehouses by 2003 in China. But its sales declined greatly in the last two years. An interesting phenomenon found in my research is that if Metro and Carrefour are in the same Chinese city, more customers like to choose Carrefour for shopping rather than Metro. A significant phenomenon is that in recent years, some warehouse operators transfer their warehouses to hypermarkets, such as Wal-Mart changed two of its Sam's Club to Supercenters. Meanwhile, the warehouse model also faces direct competition from Chinese traditional markets and the traditional wholesales market. Its membership system seems unsuitable to Chinese conditions; thus customer loyalty is hard to build. The format of warehouse is in hard times in China.

General Merchandise Store (GMS). The first GMS in China was launched by Vanguard in Shenzhen during its competition with Carrefour and Wal-Mart. It combines the features of DS and supermarket together and successfully realizes localization adapting to the local market. Compared with hypermarket and DS, GMS has advantages in both breadth and depth of merchandise, and its price range is wider. So it attracts a wider segment of customers. Vanguard is a typical example of this format. GMS is very promising in China, but it requires a high level of management, particularly in standardization and central sourcing. Japanese Ito-Yokado also operates GMS and has opened four stores in China, two in Beijing and two in Chengdu.

Specialty Store. In recent years, Chinese specialty store developed quickly. Chinese retailers dominate home electronic appliances while foreign operators have advantages in the furniture field. This format has good prospects in China. Shenzhen Golden Seahorse Group is the largest operator in furniture. Its sales were RMB 6.8 billion Yuan in 2003. Beijing Gome Electronic Appliance Company Ltd, Suning Electronic Group and Sanlian Commercial Corporation are the largest home appliance specialty stores. Their retail sales in 2003 were about RMB 18 billion Yuan, RMB 12.3 billion Yuan and RMB 10.6 billion Yuan respectively. The foreign furniture operator, IKEA, opened only two stores in Beijing and Shanghai with one store in each city. B & Q entered China in 2001; it plans to open 100 stores by 2006 cooperating with Haier.

Shopping mall. This kind of outlet mainly concentrates in Beijing, Shanghai and Shenzhen. It is becoming popular but the number may be limited. The first one was opened in Shanghai in 2000. There may be about 10 shopping malls by 2005 in the three cities.

6.3.4 The ownership structure of Chinese retailing

The ownership structure of Chinese retailing is diversifying. The monopoly of the state-owned retailers (SORs) has been broken. The state-owned, collective-owned, private-owned and JV retail stores share the industry together. The SOR's dominant position has declined while private-owned retailers (PORs) and JV stores are booming. In 2001, the Main PORs took 44 per cent of the national sales compared with only 18 per cent of the Main SOR's. Main retailers are the retailers whose sales were over RMB 5 million Yuan and swith at least 60 employees; their sales accounted for 80 per cent of the national sales by 8,425 retailers in 2001.

6.3.5 Other characteristics of Chinese retailing

Chinese retailing is in a revolutionary period. Many new retail formats have emerged and are spreading to the industry; Chinese traditional retail formats are evolving into modern formats. This evolution is faster than most people realize. Meanwhile, supported by the Chinese government, chain operation has become a main trend in the industry. The total chain companies had increased from 150 by 1994 to 2,100 by 2001 with stores increasing from 2,500 to 32,000. The largest retailer in China, Brilliance, is also the largest chain operator (see Appendix B). In 2003, Chinese chain stores' sales took 15 per cent of the national retail sales. The total sales of the top 100 largest Chinese chain retailers reached RMB 358 billion Yuan in 2003 with 20,424 stores, increased by 45 per cent than 2002. In recent years, there is a trend to be big in the industry: "the bigger, the better" is becoming popular in China. Chinese retailers are trying their best for expansion. In addition, MNRs are shaping Chinese retailing. Although their sales account for less than 10 per cent of national sales, their growth is accelerated further by China's accession to the WTO. A new industrial structure is forming.

6.4 E-tailing in China

E-tailing in China is still undeveloped. Fewer than 50 per cent of Chinese retailers had e-commerce capability. The main reasons are: "Chinese low credit card penetration and absence of a reliable banking system supporting the Internet transaction." Chinese undeveloped distribution infrastructure and chaotic distribution system. "The lack of relative regulations on e-commerce." The less know-how on modern retailing management. In addition, brick-and-mortar shops also have an entertainment function, which is a main advantage over E-tailers.

6.5 Summary

The current Chinese retail market is mainly a market with a 495 million population rather than a 1.3 billion population. But it has great potential due to the growth of Chinese GDP, the increase of Chinese population and the future urbanization of Chinese rural areas. The development of Chinese retailing is quite imbalanced. The industry presents a typical dual-structure: the developed urban retail market and the less developed rural market coexist; and modern retail formats and traditional markets coexist. Although all modern retail formats have emerged in China, the industry is still very fragmented and it is traditional retail formats that dominate the whole industry.

Chinese consumer revolutions

7.1 Chinese consumer behavior in the transitional economy

Consumer behavior is the activities that people undertake when obtaining, consuming and disposing of products and services; it closely relates to issues such as consumers' culture background, income, ethnicity, etc. Before the economic reform, Chinese consumer behavior was greatly constrained by their low incomes and the shortage of products. After the economic reform, Chinese social and economic changes, such as the emerging of middle class (M Class), the enlarging gaps in income among people, the influence of foreign cultures, make Chinese consumer behavior diversified. Generally speaking, Chinese consumer behavior presents at least the following characteristics:

7.2 Chinese consumption level is still low

Chinese consumption level is still low and mainly focuses on basic consumption, particularly on food. Although Chinese consumption keeps increasing, such as by the NBSC, the annual increase of the consumption was about 7.3 per cent from 1979 to 2003, the spending mainly focuses on basic consumption; in 2003, the Engel Coefficient for Chinese urban and rural families stood at 37.1 per cent and 45.6 per cent respectively. The high Engel Coefficient indicates Chinese life level is still low and nearly half of their spending is on daily necessities, especially food. In the next decade, Chinese Engel Coefficient will still keep high due to the large amount of rural population. Relative low purchasing power makes people very price sensitive; they are value-oriented consumers. Meanwhile, the feeling of future uncertainty caused by the reform and Chinese traditional culture in consumption make people cautious in their spending. The increasing gap between the saving deposits and national retail sales indicates that they are paying more attention to saving money. Meanwhile, influenced by the traditional culture, Chinese still like spending the cash that they have made; credit consumption has not been popular yet.

7.3 One-stop shopping is emerging in China

"One-stop shopping" is emerging in China, but most people still like shopping in traditional wet markets. With the introduction of hypermarkets and the emergence of general merchandise stores (GMS), one-stop-shopping is becoming a trend in China. Hypermarket and GSM are gradually becoming the images of "one-stop-shopping" by providing customers with a wide range of goods selections. According to Mr. Kosta Conomos, a senior officer of AC Nielson China, 87 per cent and 60 per cent of urban Chinese have bought foods and daily consumer goods in hypermarkets separately, compared with 85 per cent and 46 per cent in supermarkets. But because of the lack of enough fresh food and normally far locations, hypermarkets and GMS are only mainly places for weekend shopping. Since Chinese Consumers' mobility is low and most Chinese families depend on public transportation for their shopping, people like shopping in downtown and suburban areas with convenient transportation. Chinese consumers normally consider "quality and freshness" of commodities are the most important issues in choosing stores, particularly for buying food; the traditional wet market thus is always regarded as the main place to buy food, particularly fresh food.

7.4 Chinese consumer behavior tends to diversify

Chinese consumer behavior tends to diversify, which is mainly presented by the emergence of different consumer groups. Chinese economic development causes many social changes. The emerging and fast growing M Class are becoming a main consumption group in China. Meanwhile, due to the "One Child Policy," two large social groups are emerging: the single child generation (S-generation) and the aging group (A Group). The largest consumption group actually is Ordinary People Class (O Class). There also exists a low income class (L Class). Thus, Chinese consumers can be roughly divided into the five main consumption groups: M Class, O Class, S Generation, A Group and L Class . Each group has its typical consumer behavior (Figure 7.1).

S Generation. Since China took "One Child Policy" since 1979, the population of S generation has reached 400 million, one third of which is in urban areas. Although most of them do not have income, they were once called "little emperors" with "six wallets," which means they may obtain money from their two parents and four grandparents. In urban families, about 50–70 per cent of the total expenditures are dedicated

Figure 7.1 The market size of different segments in Chinese consumption

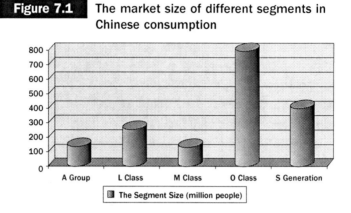

to the single child (Li, 1998). They like new fashion, new products, famous brands and new lifestyles; and their influence on their families' spending greatly (Lu, 2002).

M Class. There are many different definitions on Chinese M Class in recent years. But all of them only use one specific figure to define it and none of them considers the concept has different meaning to rural residents and urban residents and changes with time. M Class here is defined as those Chinese whose annual disposable income or annual revenue is at least twice of the national average level. In urban areas, the current M Class' annual disposable income is at least RMB 14,000 Yuan per capita or the annual revenue is at least RMB 16,000 Yuan per capita; while in rural areas, the current requirement for M Class is that the annual net income is over RMB 5,000 Yuan per capita. Then the current population of the M Class is about 130 million; and most of them live in the three areas: the Zhujiang River Delta centered at Guangzhou, the Yangtze River Delta centered at Shanghai and the Bohai Sea Circle area centered at Beijing. The M Class people are concerned with life quality and enjoying life; many of them are able to afford cars.

A Group. It refers to those Chinese whose age is over 65. According to the Minister of the Ministry of Civil Affairs, Mr. Li Xueju, China is entering an aging society; the population of A Group is nearly 134 million, and its annual growth rate is 3.2 per cent. Most of them still hold Chinese traditional consumption values and are cautious in their spending; but they tend to be concerned with quality issues in their shopping, although they are some price sensitive. Their demand for convenience is growing quickly.

L Class. It is defined as those Chinese whose disposable income or revenue is less than the half of the national average level. To the urban residents, the L Class' annual disposable income is less than RMB 3,500 Yuan per capita; or annual revenue is less than RMB 5,000 Yuan per capita. While to rural residents, the current L Class's net revenue is less than RMB 1,300 Yuan per capita. Most of the L Class live in the Central and Western China. Its population is about 250 million. They are the most price sensitive group and the functionality oriented consumers.

O Class. Ordinary People Class is the largest group by population. It refers to those whose annual revenue is between the national average level and the twice of the national level. In the urban areas, the current O Class' annual disposable income is between RMB 7,000 Yuan and RMB 14,000 Yuan per capita or the annual revenue is between RMB 10,000 Yuan and RMB 20,000 Yuan per capita; while in rural areas, the current requirement for Chinese O Class is that the annual net income is between RMB 2,500 Yuan and RMB 5,000 Yuan per capita. The current population of the O Class is about 800 million. They are in the transition from function oriented consumers to quality oriented consumers and value for money kind group (Figure 7.2).

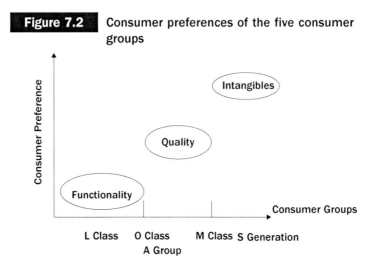

Figure 7.2 Consumer preferences of the five consumer groups

Meanwhile, Chinese consumer behavior changes by regions for its diversified cultures. For example, in Guangzhou, more consumers visit convenience stores than people in other cities; in Shanghai, consumers like modern retail formats more than traditional formats and the hypermarket is the most favorable place for daily shopping; and in Beijing, shopping in the hypermarket and department store is an important activity at weekend.

7.2 Summary

In the Chinese transitional economy, Chinese consumption level is still low and the consumption mainly focuses on basic consumption. However, Chinese consumer behavior is becoming diversified; the emergence of M Class, S generation, O Class and A Group presents four main consumption groups. Modern retail formats, such as the hypermarket, are shaping Chinese consumer behavior. Due to the Chinese "One Child Policy," the S Generation and A Group have become two of the most important consumption segments.

Part III
Competing Chinese Retailing

Entering the world's largest emerging market

8.1 MNRs in China

It has been over 10 years since China formally opened its retailing to foreign investors. Many MNRs have been in China (Table 8.1). Overseas retailers are expanding at a dazzling pace. They have injected world-class managerial expertise into their Chinese operation, which promises a high presence in local markets. With a strong emphasis on localization and brand names, they have won increasing recognition from Chinese consumers.

According to the China General Chamber of Commerce (CGCC), MNRs' retail sales accounted for 1.46 per cent of the national retail sales in 2001; (the figure was 3 per cent if the retailers from Hong Kong, Macao and Taiwan were included); in Beijing and Shanghai, the figures were 6.2 per cent and 8.1 per cent respectively. Their performance is better than Chinese retailers. For example, in 2001, foreign JVs out-performed Chinese retailers in gross margin, net profit margin and debt ratios. In 2003, the performance of MNRs is even much better. Among the top 400 retailers in China, MNRs accounted for nearly 30 per cent of market shares. Their retail sales per square meter reached averagely RMB 2.06 Yuan, higher than RMB 1.40 Yuan of the Chinese retailers. In hypermarkets or large supermarkets,

Table 8.1   Date the first store opened in China

Retailer	Country of origin	Year
Yohan	Japan	1991
Seven-Eleven	Japan	1992
Carrefour	France	1995
Daiei	Japan	1995
Jusco	Japan	1995
Mackro	Netherlands	1996
Wal-Mart	USA	1996
Metro	Germany	1996
Ahold	Netherlands	1997
Lotus	Thailand	1997
Auchan	France	1998
Ito-Yokado	Japan	1998
IKEA	Sweden	1998
B & Q	UK	1999
OBI	Germany	2000
Tesco	UK	2004

MNRs dominate the Chinese market. Among the 75 hypermarkets whose area is over 10,000 square meters per store, foreign retailers took 47 and the retail sales accounted for 74 per cent of the total sales of the 75 stores. Among the 299 stores whose area is over 5,000 square meters per store, MNRs took 157 and the retail sales accounted for 72 per cent.

Foreign retail investors entered China by the following turns in retail format: specialty store, department store, supermarket, hypermarket, convenience store and discount store. The earliest retail business in which foreign investors were involved was specialty store in the 1980s. When Chinese retailing was opened in 1992, the first entrants were from Hong Kong, Japan and other Asian countries, and their main format was the department store. The next round of entry was led by Carrefour from 1995, and most of the entrants are European or American retail giants. The main formats taken are supermarket, hypermarket and warehouse, particularly hypermarket; these formats are more successful than the department store. From 2002, foreign convenience store operators such as 7-11 and discount store operators such as Carrefour-owned Dia started to enter China; and they may be the next round of popular formats for the next few years.

8.2 The entry mode strategies of MNRs

The earliest entrants in Chinese retailing were Asian manufacturers. They normally took either a real estate based model, such as renting counters from local department stores, or a manufacturing based model, such as cooperating with their Chinese partners to open specialty stores, selling some of the products made in their Chinese JVs. But they were both production enterprises. Their main goal was to pursue production activities in China rather than retail business, and retail was just a small part of their business. So in a strict sense, they are not real retailers.

MNRs normally take one of two entry mode strategies to enter China: the direct entry mode strategy and the indirect entry mode strategy. The direct entry mode strategy includes two ways: the formal entry through the central government, which is to follow the regulations of Chinese retailing and to develop step by step, and the informal local entry, by which MNRs directly enter a Chinese local market with the permission from the local government rather than the central government. By the first way, approved by the central government, MNRs can obtain trading rights, facility of expansion and airtight legitimacy. By trading rights, foreign investors can import goods for selling in their own stores and procure Chinese goods for export to balance their foreign exchange. Because these kinds of JVs operate under clear published rules, they are unlikely to be delegitimized. Actually, many MNRs indicate that they normally do not like to build JVs with Chinese retailers. They often choose Chinese partners without any retail experience, by which MNRs meet the request of the central government while independently managing the retail business like a wholly-owned company, because the Chinese JV partners they chosen often lack experience and ability to manage the retail business. So the intention of the Chinese government to make Chinese retailers MNRs' partners and learn management know-how in their cooperation has not been well realized. The second way, local level entry, is to obtain permission from Chinese local governments, which is actually illegal. By 2000, the number of central government approved retail JVs was 28 while the local approved was nearly 300. One popular way for local governments to approve JVs was to permit foreign investors to venture with Chinese companies who had already had a retail license; and the new JV just used the exiting retail license. Many Asia-based investors built their retail ventures quickly in this way. The central government noticed this phenomenon and began to clear up this kind of JV in 1997. The indirect way is a way of accomplishing things that adheres to all laws and regulations, but is not what the regulators had in mind when they grew up. By this way, MNRs normally take indirect investment, involving contractual payment and service rather than equity. The main indirect entry mode strategy is the management contract way, by which a foreign investor has no equity in Chinese JV but manages the retail company.

The management contract normally grants that the foreign partner takes complete management control in return for a management fee from the Chinese party. The foreign partner usually provides training and equipment and keeps some measure of control by holding the store's trademark. Although this way is useful in helping MNRs to understand Chinese retailing and local markets, it is risky from both the operation and the conflict with retailing regulations. There have been cases in which the Chinese partner has run off with the foreign investor's marketing formula. So if a foreign investor choose this way, it is very important to choose a reliable Chinese partner.

8.3 Analysis of the difficulties for MNRs operating China

Theoretical analysis: Transition, transaction cost and transition cost: The first difficulty that MNRs meet in China is that China is a very different market. The difference is not only from the different consumer behaviors but also from the different rules generated from the transitional economy, which often makes MNRs' CAs and successful models lose their edges.

The transaction cost in the Chinese transitional economy is very high. A company's transaction cost normally comes from two sources: the economic system that the company operates in and the company's own operation model. A different economic system defines different transaction cost. As a transitional economy moves forward and the progress of its marketization, the transaction cost from the economic system or the transaction cost of the whole economy tends to decline. When an economy changes from one economic system to another, such as from a planned economy to a free market economy, a transition cost is generated. The transition cost can be defined as the difference of the transaction costs between the two economic systems or between the two economies. Transition cost also can be defined as a narrower concept as the industrial transition cost, which refers to the transaction cost difference of an industry between two different economies; or as a company's transition cost, which refers to the transaction cost difference when the company operates in two different economic systems. For example, in the case of retailing, the transaction cost of retailing in the transitional economy is higher than that in the free market economy. But both of them tend to decline with the development of economy that they exist in. The difference of the transaction costs between the two retailing industries is the transition cost of retailing. It can be argued that the process of an economic transition actually is the process of making the transition cost diminish or the process of reducing transaction cost.

Transition cost mainly comes from two sources: the economic system itself and institution arrangements in the economic system. The former defines the nature of competition while the latter defines the degree of competition. The transaction costs between two free market economies are also different; this is mainly because of different institutional arrangements between the economies. Different institutional arrangement results in different ways in administering corporate operation and market supervision and then generates different transaction costs. When a retailer enters the Chinese transitional economy from a market economy, it has to bear the transition cost in its operation. This transition cost comes from the change of economic systems and its individual extra cost to develop the new business model and CAs. The former can be further divided into two costs: the retailing transition cost and the non-retailing transition costs.

8.4 Practical analysis of the main difficulties MNRs meet in the Chinese market

In reality, competing in the Chinese retailing is hard. MNRs often encounter kinds of difficulties, these mainly include:

The shortage of management personnel in retailing: There are serious shortages of management skills and expertise in China, particularly in store management, and purchasing and inventory management, which hold back MNRs' development. It takes a long time to train qualified managers while the transfer of qualified managers from abroad is costly and may not be effective as they may find it hard to adapt their experience to local markets.

Good location for store opening is often hard to obtain: In retailing, the golden rule is location, location, location. However, good location is a scarce resource and has often been taken by local retailers. In Guangzhou, Carrefour planned to open eight hypermarkets since 1998. However, five years have gone, and it opened only one because it could not find ideal places to open them. Similarly, in Guangzhou Wal-Mart took two years to find an ideal location. Two reasons can explain this. One is that ideal locations are limited. Both Wal-Mart and Carrefour ask that the place should be about 20,000 to 40,000 square meters in size, near main roads with convenient transport, and should have enough rich residents' districts nearby. The available locations meeting these requirements are rare. The other reason is the price of the location. Good location means expensive price. MNRs normally restrict their rent cost to between 2–4 per cent of their profit, while local landlords often want to make a lot of money from MNRs, because they think foreign investors are able to afford high rents. If the rent is too expensive, MNRs have to give it up.

Tax problems: In China, tax includes local tax and central tax. Each store must pay them separately according to its location. For a retailer with many chain stores, each store has to pay tax by itself, which makes each store become an independent operator. The central payment of tax is impossible in most Chinese cities except Shanghai. In Shanghai, the headquarters of a chain store operator is allowed to pay tax for all its chain stores; then the government allocates 70 per cent of the tax to local/district governments.

Troublesome procedures for opening a store: In China, if a retailer wants to open a store, it needs various kinds of operation licenses including a registration certificate, hygiene certificate, wine and tobacco license, license for newspapers and magazines, business license, statistics certificate, etc. Retailers have to go to different government departments for different procedures. The application for them is not only cumbersome but also time consuming (Table 8.1). It takes a retailer at least 90 days to obtain all the paperwork for opening a store; and the same set of materials must be photocopied at least 10 times. For example, a Shenzhen based retailer with 109 stores needs to apply for 2,180 licenses for all of its stores. Because it takes so much time, 80 per cent of retailers in Shanghai opened their stores while they were applying for the licenses even though it is illegal to do so.

Regional protectionism: This not only makes it difficult for MNRs to build a distribution system crossing regions but also makes it costly to develop local markets because of some discrimination policies. For example, Shenyang even wanted to charge Wal-Mart a RMB 100,000 Yuan river-clean fee to open a store, which is much more than domestic retailers would be charged.

Cost of licenses for opening a store: Application for various kinds of licenses means that retailers must pay much money

Table 8.2 Some application procedures and the time taken

Procedures	Time needed
Registration	10 days
Hygiene certificate	30 days
Alcohol and tobacco license	30 days
License for newspapers and magazines	10 days
Branch store code	10 days
Business license	10 days
Tax registration	15 days
Statistics certificate	10 days

for them. To open a store, these fees are about RMB 1,000 Yuan. In addition, there is an annual inspection fee of RMB 500 Yuan for each store. It may not seem too much money, but for a retailer with a chain operation, the money could be considerable; for the retailer mentioned with 109 stores, the amount of money paid for the applications is over RMB 160, 000 Yuan. Further, sometimes there are extra fees in different regions. Wal-Mart was asked to pay river-clean fees of RMB 100,000 Yuan, about 0.1–0.2 per cent of its sales.

Difficulty in standardization of operation: The fragmentation of Chinese regional markets, characterized by culture diversity, makes harmonization of product ranges difficult, especially in food where each region has its own traditions and tastes.

Legislation environment is often worried: When China had just entered the WTO, it allowed just 4 JVs in Beijing and Shanghai respectively and no more than 2 JVs in other cities. But the reality is that more JVs are built there. In Chongqing, there should be no more than 2 JVs; actually Carrefour, Metro, Parkson, Pacific Department store had opened stores there. Their policy is that "we just do it, but do not speak out what we are doing". The complicated regulation situation brings difficulties that MNRs have never met before. Success in China is not easy.

Cost in market supervision: In China, market supervision and research are very hard for reliable data are normally difficult to obtain.

8.5 The impacts of MNR entry on Chinese retailing

Foreign investors had built 350 retail companies by 2003. The entry of MNRs creates what Schumpeter called a "big disturbance" in Chinese retailing, because they disrupt the existing system and enforce a distinct process of adaptation. For example, the entry of MNRs, such as Carrefour and Wal-Mart, caused disruptive impacts on Chinese retail price and wide selections of goods. The disruptions are presented by the dramatic change of the Chinese retailing structure, particularly the retail property system and retail format structure. Since MNRs entered China, the monopoly of SORs has been broken down:

1. From the perspective of property rights, before MNRs entered China, it was the SORs that dominated Chinese retailing. By the format of JV, MNRs not only diversify the property rights system but also encourage the development of non-SOEs. Most suppliers of MNRs are private-owned enterprises, which grow alongside MNRs. Both MNRs and their private-owned suppliers are often very competitive companies in Chinese retailing. They have a great influence on the Chinese economy.

2. MNRs diversify Chinese retail format by bringing modern retail formats, such as hypermarkets introduced by Wal-Mart and Carrefour; convenience stores by Lawson and 7-11; warehouses by Metro and Makro, etc. These retail

formats, which are market-oriented, have become consumers' top choices. The traditional department store dominating Chinese retail market has been weakened greatly.

3. From the competition perspective, the entry of MNRs brings the "catfish effect," resulting in real competition in Chinese retailing. Foreign participation gives a strong stimulus for developing Chinese retailing. The competition from MNRs has forced Chinese retailers to upgrade their management and service with the latest technology and to speed up its consolidation. The sector is setting up some large brand-name retail stores through M & As and alliances supported by local governments so that Chinese firms are able to compete with their international rivals. On the other hand, an aggressive influx of MNRs inevitably gives a blow to domestic retailers. Their reduced market share in large cities is expected to shrink further. SMRs are the hardest hit and some of them are going under.

The impacts of the foreign entry on Chinese retailing is just as Huang Hai said, the economist and director of the Bureau of the Internal Trade and Market at the SETC, "This foreign commercial presence in China goes some way to promote the development of the industry and the national economy. Chinese retailing has being shaped by MNRs and different retail formats. All Chinese new formats are copied from foreigners." The penetration of foreign retail formats has given a strong boost to retail sales in the Chinese market. MNRs not only bring international capital but also advanced management technologies, retail format, international marketing methods and management know-how, by which they educate Chinese retailers. And the most important is that they bring consumers new lifestyle and new consumer concepts, which will influence Chinese retailing in the long run. Meanwhile, the early entry of MNRs also benefits themselves. They not only achieve FMA but also obtain good sites and favorable rents for their businesses. They can develop duplicated models or experiences earlier than later entrants for their further expansion. By actively localizing their businesses, not only can MNRs better respond to local customer demands but also reduce their purchasing costs and build complete supply systems. Relying on their global outsourcing networks, foreign-funded retailers generally have a marked competitive edge over their local rivals in terms of commodity prices. However, the entry of MNRs also brings some bad influence in Chinese retailing. Some MNRs do not obey Chinese laws and regulations in their operation: they sell goods beneath cost price and transfer their operation costs to suppliers, which seriously breaks the rules of competition and causes unfair competition. Carrefour could be called a "bad" competitor for these reasons (see the case of Carrefour).

8.6 Summary

Since China formally opened its retailing to foreign investors, many MNRs have been in China either by the direct entry mode strategy, or the indirect entry mode strategy. They entered China by the following turns in retail format: specialty store, department store, supermarket, hypermarket, convenience store and discount store. The entry of MNRs creates what Schumpeter called "big disturbance". Many of them out-performed Chinese retailers in gross margin, net profit margin and debt ratios; and these MNRs presence in China goes some way to promote the development of the industry and the national economy. But their operation in China is also facing challenges because of Chinese conditions.

The success and failure of global retailers in China

9.1 Can MNRs succeed in China?

Not all MNRs can harvest successful fruits in this, the world's largest emerging market. When some of them are enjoying the success, others are experiencing failure. In the 1990s many MNRs rushed to compete in Chinese retailing while many of them did not have enough knowledge of the market. So they could not successfully develop appropriate operation models and encountered both internal and external obstacles in their operation. How to succeed in Chinese retailing? Some failure lessons may give some answers.

9.1.1 Localization is a basic rule for surviving the competition

Retailing is often called a local business. In a retailer's internationalization, localization is the most important and also the most difficult. Without successful localization, the success of the internationalization is impossible. In the localization, the retailer must understand its local consumer demands well so that it can meet the demands better than its competitors. In a large market such as China, localization is particularly difficult, because there are many sub-cultures in different regions within Chinese culture, which indicates that local tastes are greatly diversified. The research undertaken by EUI showed that 44 per cent of the surveyed companies treated China as one market; 6 per cent treated China as two markets; 11 per cent treated China as three markets, and 39 per cent treated China as over four markets (Figure 9.1). The case of Park' N Shop, a Hong Kong based retailer, might give an example of the difficulties of localization.

The Case of Park' N Shop. Park' N Shop is a Hong Kong based retailer under Hutchison Whampoa. It is one of the largest retailers in Hong Kong operating over 200 SM supermarkets there. In 1984, it entered the neighboring region, Guangdong Province of Mainland China. Because people in Hong Kong and Guangdong share a similar culture and lifestyle, consumer behaviors between them are quite similar. Therefore, Park' N Shop targeted Guangdong and thought it would be easier to succeed there than in other markets. It developed quickly there at first and the store number even reached 52 including 11 in Guangzhou, the provincial capital. However, because it could not successfully realize localization, the company lost money seriously for over 10 years. In 2000, Park' N Shop had to close its 11 stores in Guangzhou and 4 stores in other cities.

Figure 9.1   How foreign companies treat China as they develop the market

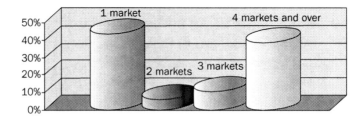

The reason for its failure is because it only copied its Hong Kong model to Mainland China:

- *Retail format*: It is a 500–800 square meters supermarket

- *Product assortments*: Most products were imported with very expensive prices. There was no fresh food.

- *Operation management*: 60 per cent of employees were from Hong Kong, which not only caused high cost but also poor understanding of local consumers.

In 1999, Mr. Iwan Evans, who came from Wales in the UK and had 20 years work experience in retailing, was appointed as the General Manager of the China District. He asked for repositioning of the stores and advocated complete localization. Hong Kong based employees were cut from 60 per cent to 3 per cent in order to employ as many local people as possible. "Only local people can understand local consumer demands well," Mr. Iwan Evans said.

- *Target mass-market, especially the female segment*: The stores adopt a low price image and claim "Let women of the whole city disappear into our stores."

- *Focus on fresh foods*: Mr. Iwan Evans took the management team to do market research and found local people like fresh food. He said his store must sell fresh food that people use every day such as live fish.

- *Retail format*: Open a large supermarket that is several thousand square meters.

Then at the end of 2000, 18 months after Park' N Shop withdrew from Guangzhou, it came back again with a large fresh food supermarket, Zhonglu branch. The store is like a traditional wet market just focusing on fresh food at low price. Therefore, it achieved great success. By 2001, the sales of the store kept increasing by 6 per cent to 8 per cent per week. Thus, the second generation Park' N Shop store was born. Six months later, in July 2001, a larger store, a 13,900 square meter hypermarket, Jintian branch, opened. The hypermarket focuses on fresh food and low price merchandise. Fresh food accounts for over 50 per cent of the total assortments. This store is so successful that it has become the sign of the third generation of Park' N Shop. Iwan Evans said "Besides meeting local tastes, fresh food is a kind of daily necessity, so the demand is huge. Although the profit margin is low, we can succeed by huge quantities." The success greatly encourages the company. Park' N Shop plans to invest over RMB 100 million Yuan in developing the Mainland market using the similar model. The successful localization model of Park' N Shop is very simple: fresh food + low price + hypermarket. But the cost to understand this is quite high. The core of the model is to deliver what local consumers really want at the price they like to pay. From the Park' N Shop case, the importance of localization can be seen; even in two regions with similar culture and lifestyles, localization is still needed, which still is a key issue for retail success.

9.1.2 Choosing the right retail format is the key to retail success

The case of Yaohan. Japanese based Yaohan announced its withdrawal from China in August 1998, because it could not see its supermarket chain in Shanghai growing as expected while many hypermarket operators, such as Lotus, Auchan and RT-Mart had succeeded there. Yaohan took the format of supermarket in Shanghai, which requires a large number of managers who not only understand both Chinese retailing and their own corporate strategy well but also can combine them together. This type of manager is expensive. So it had to take a high price strategy, which made a lot of customers go away. However, the larger hypermarket format easily achieves economies of scale and just needs a few bilingual managers who can interact with corporate personnel; therefore the operation cost is low. Choosing the wrong retail format was the main reason for Yaohan's failure. Actually, behind the retail format it is the target market. Different retail formats have different advantages targeting different segment markets. If the wrong retail format is chosen, it often results in the wrong target market and developing the wrong CAs. Yaohan chose the supermarket format, which indicated it should target the mass market but it developed a high price strategy, which was a disadvantage when competing for the mass market.

9.1.3 Developing CAs is important for survival

The case of Royal. Ahold Netherlands based Ahold entered China in 1996 by the takeover of 16 supermarkets from Zhonghui Non-Staple Food Corporation in Shanghai. It then introduced its international management techniques and much capital into the supermarket chain. Given the shrinking profit margin in retailing, caused by increasingly intensive competition, Ahold soon found: (1) It invested too much money in a short time. To make a profit, it had to keep a high price level, which was higher than local consumers wanted to pay. (2) It could not develop CAs compared with its competitors. Local chain operators provided similar commodities with much cheaper prices, while Ahold failed to develop either cost advantage or differentiation advantage; thus local cheaper chains were more competitive. Therefore, three years later, in 1999, Ahold had to hand control of the chain back to its Chinese partner and left China. Failure to develop CA during the competition is the main reason for Ahold's failure.

9.1.4 Developing the right strategy is necessary for retail success

To succeed, MNRs should perform their advantages fully while they attack their competitors' weak aspects. The main disadvantage of Chinese retailers is that they lack enough finance funding for their expansion while strong financial strength is just one of MNRs' advantages. Therefore, MNRs may take the following strategies for expansion and competition:

- Control some local retailers by taking their majority shares;
- Buy out the sales rights of some popular goods in some regions and cut off the supply of those products to Chinese retailers;
- Become involved in making the rules of Chinese retailing, such as shortening payment term for suppliers.

9.1.5 Building effective distribution channels to support retail success

In retailing, distribution is very important. Developing efficient distribution systems is essential to success. The distribution channels available in any country are the result of culture and tradition (Jain, 1996). Retailers must understand local conditions and examine the existing distribution system carefully to choose and build the best system within the market (Keegam, 1999). Using one standardized distribution system for international operation in every country is dangerous. That's one main reason why Wal-Mart did not copy its distribution system when it entered China.

9.1.6 Developing flexible strategy is important

Keeping their operation flexible to adapt to different local markets is important during a retailer's internationalization, particularly to a large market like China, with many different sub-cultures. A successful global model should focus on local adaptation and partnerships. Some MNRs failed in China, just because they were too successful in their home markets and continued doing what they had done. Their successful models developed in their home markets often do not suit the Chinese market, and keeping the former models rigid makes them fail to realize localization and only results in failure.

Besides the above lessons, the experiences of Wal-Mart and Carrefour also provide insight on how to develop the Chinese retail market.

9.2 Competing in China: the case of Carrefour

A brief introduction to Carrefour. Carrefour is the most international retailer in the world. It was created in 1959 in France. In 1963, it invented the concept of the hypermarket. Since 1973 it has had to pursue internationalization because of the strict legislation for developing hypermarkets in France. In 1999, 11 weeks after Wal-Mart announced its acquisition of Asda, the British third largest retailer, Carrefour merged with Promodes to defend against acquisition by Wal-Mart, which made Carrefour the largest retailer in Europe and the second largest in the world. Carrefour has three main retail formats: hypermarkets, supermarkets and hard discount store. It also has convenience stores and cash and carry stores as the complementary formats. But it mainly uses the hypermarket for its internationalization. In 2003 its total sales reached LL70.49 billion in 30 countries with 6067 stores and over 50 per cent of the sales came from its international market, particularly from the emerging market. It makes Latin America and Asia the engines for its future growth. Its corporate strategy is to try expanding its presence in local markets by each of its formats.

The entry mode strategy of Carrefour. Carrefour always regarded China as its engine for long-term corporate growth. The situation of Chinese retailing in the 1990s provided Carrefour with a great opportunity to introduce its hypermarket to China, because it would face no competitor in the whole country, which could make Carrefour benefit fully from the FMA; more importantly, the hypermarket format could gain market share quickly. This could make Carrefour fully enjoy the fast growth of the industry and the considerable industry profit margin in the 1990s. Carrefour believed that the only obstacle to its developing the Chinese market was from the industry policy: the entry must be permitted by the central government and must take the JV mode in 11 selected cities. After succeeding in Taiwan, Carrefour had gained some experience in understanding Chinese culture. Further, it had developed a management team that could be transferred to Mainland China. Then in December 1995, it entered the market as a managing consulting company to open stores, which was an unusual way to enter the market. This indirect entry mode strategy involved high risk for it broke the industrial regulation. If anything, Wal-Mart has been held back in China not by lack of demand but by getting central government approval for its stores. Carrefour, meanwhile, ploughed ahead without them.

Carrefour built a JV with Zhongchuang Commercial Company in Beijing, which was called Jiachuang Commercial Management Advice Service Corporation. By Chinese law, a commercial management company can only provide consulting services and is not allowed to do investment business. So the JV did not qualify to do Chinese retail business. Then Zhongchuang Commercial Company registered a wholly-owned shell company, Chuangyijia Commercial Company, for specially developing retail business such as opening hypermarkets. Because Chuangyijia was a Chinese company and registered for retail business, it could do any retail business without the policy limitation. Then Chuangyijia asked the JV, Jiachuang Commercial Management Advice Service Company, to manage the hypermarket, by which Carrefour became involved in the retail business; the hypermarket was named Carrefour, although it was registered by the name of Chuangyijia, which was illegal in operation, because commercial management was not allowed to do business involving its brand transfer (Figure 9.2).

In this way, Carrefour avoided the policy restrictions of applying for permission from the central government. Then, its first hypermarket was opened in Beijing; it entered Chinese retailing through this kind of indirect way. In the same way, Carrefour opened four hypermarkets in Beijing. All of them achieved great success, but none of them was registered by the name of Carrefour, although all names of stores presented were Carrefour. After that, Carrefour

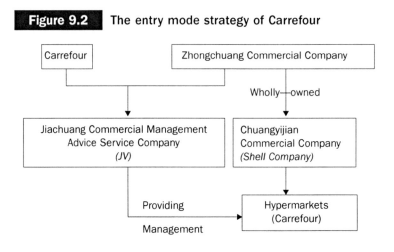

Figure 9.2 The entry mode strategy of Carrefour

quickly expanded to other cities, such as Shanghai, Tianjin, Shenzhen and Chongqing. By 2001, Carrefour had built 27 stores in 15 cities. However, none of the 27 stores were approved by the State Council. Instead, it just applied for permission from local governments, which was illegal.

Further, all the stores were directly invested and wholly-owned by Carrefour, which seriously broke Chinese retail regulation that a foreign retail project must take JV and the Chinese partner must take the majority stakes in the JV. Therefore, all Carrefour's stores were illegal. But Carrefour did not care about this and was hungry to acquire market share. In 2001 its sales in China were about L1.341 billion, a 15 per cent increase on that of 2000. Since then, it has become the largest foreign retailer and the second largest retailer in China. Because it firmly obeyed the regulation, Wal-Mart opened only 8 stores within five years of its entry. But it concentrated on procurement in China to support its global outlets. Even though the illegal status of Carrefour in China was uncovered by newspapers, it still had not applied for the license from the central government. Carrefour's entry mode strategy worked initially in its favor allowing it to beat Wal-Mart to the secondtier cities, but its illegal operation resulted in its punishment later.

Carrefour's growth strategy. Carrefour mainly takes two ways for expansion in its global internationalization: acquisition and franchising. When entering emerging markets, it often takes acquisition and chooses to retain those local retailers' names that have strong brand name recognition in the markets, such as GS in Italy and Norte in Argentina. While entering mature markets with strict regulations, it often takes franchising. By 2001, it had about 4,000 franchise stores operating under Carrefour banners. However when Carrefour entered China, it took the wholly-owned ventures. Armed with its rich international experience, Carrefour took an aggressive "multi-point entry" strategy for expansion, which was to enter several cities at the same time. In this way, it opened hypermarkets in Beijing, Shanghai, Shenzhen, etc. and established its geographic perimeter in very short time. It divided the Chinese market into the Northern, Southern, Eastern and Central China areas and took different regional strategies in each area separately. Its principle was "opening stores wherever they can be opened" and "doing purchasing separately and locally." Each store built its own purchase system, because it did not have time to build a central purchase system and apply a central management. Carrefour has a very ambitious expansion plan to open at least 10 stores each year. Although the plan was once delayed by the investigation of its illegal status, its ambition for fast expansion has never been changed. This strategy makes the company benefit fully from its FMA. However, the operation cost is high, because economies of scale cannot be achieved; central management and standardization are difficult to realize because of different conditions in local markets.

Carrefour's aggressive expansion model. Carrefour mainly targets the M Class and the O Class and develops its Chinese market by hypermarket. In 2003, it plans to introduce its discount store to China, which will be the first discount store in China.

The expansion model: Carrefour firmly takes the speed to market principal for expansion; its expansion model is simple: *speed* plus *scale*. Speed refers to opening as many stores as possible in different regions in a short time. Because the Chinese market is so large that it will take Carrefour a long time to develop national coverage, it chooses several large cities to enter at the same time; it is not unusual to see several Carrefour stores opened on the same day in different cities, which is unlike Wal-Mart's saturation strategy, filling up one market first before expanding to another city. Scale refers to using the hypermarket format for expansion, which allows Carrefour to achieve market coverage easily and fast.

The one-person development model: When Carrefour enters a new market, it often uses its unique one-person development model for developing the market. It normally just sends one person to the market at first. That person is the regional manager of that market. His first work is to recruit local people for market research, such as what the most popular fast moving products are there; then to find their suppliers directly, by which to decide product assortments and to build supplier relationships. This makes Carrefour realize localization better than other MNRs in China. In addition, its store manager master system, which is that store managers are totally in charge of their store operation, making overall decisions for the business, gives store managers more power than other retailers in operation and makes the store respond to market change in time.

The localization of Carrefour in China. The success of a retailer's internationalization mainly depends on two key operations: if the retailer is able to develop sustainable CAs and if it can successfully realize localization. Carrefour achieves these by its unique ways.

Localization: The key issue in realizing localization is to understand local culture and local consumer demands. Carrefour achieves this by its successful one-person development model. Carrefour has achieved great progress in its localization. Its Qingdao branch, the busiest store among all its Asian stores, has been rebuilt as a three-floor store to adapt to the market, which was the first three-floor Carrefour store in the world.

The localization of promotion: Carrefour used to pay more attention to western festivals, such as Christmas than Chinese festivals; but in recent years, it has begun to be concerned more with Chinese festivals. During the Spring Festival period, more and more Chinese culture has been considered. Carrefour not only specially prepares goods for the festival but also decorates its stores according to Chinese culture, such as antithetical couplets and celebrates like Chinese families. "When in Rome, do as the Romans do," a manager of Carrefour said, "The secret of success exists in, besides good management thinking and complete management system, understanding residents' life. The reason we specially add Spring Festival Counter is to bring customers a feeling of home."

The localization of management: This is the most important and also the most difficult work in localization. At the early stage of entry, Carrefour directly transferred its management team from Taiwan, Hong Kong and France to Mainland China. But with the fast expansion going on, Carrefour felt the lack of management personnel had become its bottleneck to further expansion. To solve the problem, in 2000, Carrefour built a management school in

Shanghai for specially training store managers and developing local management teams, which was the first of this kind of school in China. "The localization of employees is a foundation for our development," said Mr. Chen Yaodong, the deputy executive director of Carrefour China.

The localization of procurements: Local products normally meet local people's taste well. In its stores, over 95 per cent of the products are sourced locally from China, particularly from the regions where the stores are opened or nearby. In each city, Carrefour provides different products adapting to different local lifestyles and cultures. Besides this, inspired by Wal-Mart, Carrefour began to pay more attention to procurement in China. Its procurement in China increases greatly year by year. Carrefour also wants to make China its global supply center, as Wal-Mart has done. In 2001, its procurement reached US$1.5 billion, 5 times its US$300 million procurement in 2000. In 2002, the target was about US$2 billion and it claimed the figure would be doubled within next three years. According to Mr. Philip Pelegru, the CEO of Global Sourcing, China has contributed to about 30 per cent of its global sourcing and 60 per cent of its Asian sourcing. China has become its largest supply center in Asia, supplying its outlets in 31 countries. Carrefour had built 10 purchase centers in China by 2002 including Shanghai, Guangzhou, Beijing, Tianjin, Dalian, Qingdao, Wuhan, Xiamen, Ningbo and Shenzhen. It is developing procurement networks in Hong Kong, Shanghai, Guangdong Province, Zhejiang Province, Fujian Province and Shandong Province.

Developing sustainable CAs

Location Advantage: In retail business, location is an important sustainable CA in competition. All Carrefour's stores are located in city centers, with the benefits of easy access by convenient public transportations and a high volume of customers. Unlike Wal-Mart and Price Smart, which also do the wholesale business besides retail in China, Carrefour only does retail business. This is why Carrefour always keeps its principle of choosing city centers to open its stores. Because it opened hypermarkets in China earlier than others, it had more favorable conditions to choose locations than later entrants.

Good relationships with local governments: Good relationships with local governments are a strong CA in China. Many local governments make importing Carrefour as an important means to modernize their local commerce. This is why Carrefour is so popular and could obtain local approvals easily at its early expansion stage, which contributed to its fast expansion in China. Its successful cooperation with local governments creates favorable conditions for its development in the areas.

Low price: Low price is Carrefour's main weapon in competition. The low price is achieved by offering a limited and highly targeted line of products. To do so, it tends to choose SM suppliers, which gives Carrefour stronger bargaining power than choosing large suppliers. Each Carrefour store has about 500 suppliers. By its large quantity purchasing, it squashes the price to the lowest level. But sometimes this can damage relationships with its suppliers.

Competition in China: Because Carrefour expands too aggressively; it causes an "earthquake effect" in the local retail market and involves intensive competition with local retailers. In most cases the competition is presented by a price war. Carrefour tries to use low prices and large selections of products to win over its local competitors. In competition, some local retailers go under, but some local retailers learn quickly and grow up fast. For example, in Shenzhen, Carrefour competes with Renrenle, a local retailer whose names means "everyone is happy" and is just 1.8 km from the Carrefour store. Carrefour claimed it would crush Renrenle within three months after it entered the market. It forced its suppliers not to supply Renrenle; otherwise Carrefour would stop purchasing from them. Then it launched a two-year price war; two years' later, Carrefour had to ask for the price war to be stopped. During the competition with Carrefour, rather than being crushed by Carrefour, Renrenle became stronger and stronger, its store grew from 2,000 square meters to 12,000 square meters, and the profit is much more than Carrefour.

Competition with Taiwan based retailers: Actually, Carrefour never thought local retailers would be able to compete with it. But the competition from local retailers has made it feel threatened. Besides local retailers, Taiwan based retailers bring Carrefour an unexpected headache. Taiwan based retailers are too similar to Carrefour in operation, such as in strategy, location selection, management and marketing. In Chinese Carrefour stores, Taiwanese account for the majority of the management teams, especially in middle manager class, and many of them changed to Taiwanese based retailers later. Many suppliers think Carrefour and Taiwanese retailers are nearly the same in the overall business operation. Carrefour has advantages in the FMA, brand, and human resource, while Taiwanese retailers have advantages in localization and marketing because of their better understanding of Chinese culture. They follow Carrefour and compete with it in where Carrefour opens stores. For example, Taiwanese retailer Hy-mart, which was invested by Tinghsin International Group in 1997 taking the format of hypermarket and shopping mall, had developed 12 stores in Mainland China by 2001. Its strategy is to directly compete with Carrefour. In Shenyang, the largest city in Northern China, Carrefour became involved in a price war with Hy-mart; Carrefour could not win the competition, because Hy-mart's store could attract many customers from the Carrefour store by its better understanding of local consumers. Since 2002, the sales of the Carrefour store in Shenyang have decreased gradually. In March 2002, Carrefour's customers were halved because of the competition from Hy-mart. In other cities, such as Dalian and Chongqing, the same story is repeated. Therefore, Taiwan based retailers have brought great threats to Carrefour.

Carrefour's problems. Although it claims that "the sales revenues in China have been extremely satisfactory with double digit growth in sales with constant comparable sales area" (Carrefour Annual Report 2001), Carrefour has many problems including some serious problems, which may prevent its further development if they cannot be solved in time.

The relationships with suppliers: Developing good relationships with suppliers is essential for retail success. But Carrefour has serious problem in this respect. The main reason that Carrefour failed in Hong Kong in 1997 was because of bad relationships with its suppliers who could not tolerate the thin profit margin and unfavorable supply conditions Carrefour asked. In 1997, its 22 suppliers in Hong Kong decided to stop supplying Carrefour products altogether. Similar cases also often happened in Mainland China. For example, in 2001, some suppliers in Chongqing city stopped supplying Carrefour products, because Carrefour dumped their products by much cheaper prices than the contract prices. Many suppliers of Carrefour complain about Carrefour's payment conditions, particularly its long-term payment (Table 6.4). By the long-term payment, Carrefour uses a lot of suppliers' capital and saves huge own money. In Chongqing city, Carrefour uses suppliers' money for at least 300 million each year. Most of its money for the procurement is actually "invested" by suppliers. Carrefour just uses suppliers' money to make money.

A supplier complained that, his first supply was about RMB 50,000 Yuan; then the second supply, the third, and so on. The more products he supplied Carrefour, the more money he "sent" to Carrefour, until he got the first payment 60 days later. Besides, suppliers must provide logistic services for Carrefour. The following is a part of the supply contract translated from the Chinese version:

"Besides guaranteeing the quality of products provided, the supplier should provide the most favorable price for Carrefour. Supplier must transport products to the store in time by the date showed in the contract; if delayed, Carrefour will charge 0.5 per cent of the total value of the delayed goods each day for punishment."

Further, Carrefour also charges "channel fees" from suppliers and makes them the main profit resource. A Chinese channel fee is a kind of transaction fee that Chinese suppliers must pay their retailers in order to have their products displayed in the retailers' stores for selling. It is

Table 9.1 Comparing average payment terms among foreign based retailers

Retailers	Payment conditions
Wal-Mart	30 days
Mackro	30 days
Price Smart	30 days
Trust-Mart	30 days
Metro	30 days
RT-Mart	30 days
Auchan	45 days
Lotus	45 days
Carrefour	60 days

similar to the slotting fee or slotting allowance existing in Western countries, which is paid by a vendor for space in the retail store, but different from it in both amount and items. In 1998, Carrefour charged 14 kinds of channel fees from its suppliers and even charges a fee for rebuilding its store. In Beijing, a supplier provided Carrefour with about RMB 200,000 Yuan of products, and got only about RMB 300 Yuan back after the channel fees were charged. Despite low prices, suppliers have to accept the 60 days payment term and provide Carrefour with logistics services, which make it hard for suppliers to survive. Therefore, Carrefour shows little consideration for the interests of suppliers, which brings a negative influence to Carrefour's operation.

The relationships between low price strategy and product quality: Carrefour wants to keep a low price image, but sometimes it is so concerned with low price that it even sacrifices product quality to it, which seriously ruins its reputation. For example, in August 2001 in Chengdu, a large city in the Southwest, Carrefour was found to sell a large number of fake CDs. In Wuhan, the largest city in Central China, customers found some bread Carrefour sold with three labels in June 2001, each of which showed a different expiry date; the bread had already become uneatable. It was also found to have sold an unsafe plug there. After customers complained about the problem, it still displayed the products on the shelf. Many similar cases have been found in different cities in recent years. In addition, a lot of customers complain of its poor service. Carrefour won the prize for the most customers' complaints in Zhejiang Province in 2002. These facts suggest that serious problems have been in Carrefour's management.

The punishment: for illegal operation: In early 2001, Carrefour was investigated by the central government for its illegal status. It was asked to stop expansion before meeting the requirements of industry regulation. Then Carrefour's fast expansion pace was stopped. It had to re-apply for permission from the central government and reorganized its retail projects in China. In the past, Carrefour always kept secret how many of its stores were wholly-owned rather than JVs. It admitted finally that all of its 27 stores were 100 per cent wholly-owned. After hard negotiation with the Chinese government, Carrefour had to agree to restructure its stores and sell 35 per cent of the shares of all 27 stores to Chinese companies in order to make its local companies JVs in early 2002 when MNRs were allowed to hold up to 65 per cent of the shares in a JV.

On June 12th 2002 the *Financial Times* reported that Carrefour would have to sell out at least 35 per cent of the shares of its three wholly-owned stores in North East China cities including Shenyang, Ha'erbin and Dalian city respectively, to Chinese companies. After "making up its procedures" Carrefour is picking up where it left off, but its business had been seriously influenced. For example, Carrefour's expansion plan of opening 10 stores each year had been delayed for over one year, during which Wal-Mart, which had gained a good reputation for obeying Chinese regulations, opened 8 stores, making Chinese stores reach a total of 19. Carrefour's case reminds MNRs that being patient and obeying national regulations in developing the Chinese market are important.

Assessing Carrefour's performance in China. In terms of market share, we can say that Carrefour is successful in China. In 2003, Carrefour's annual sales in China reached L 1.45 billion, ranking the sixth largest retailer and the largest foreign retailer in China; and China has been the Carrefour's leading Asian performer in terms of sales. While if we compare Carrefour with Wal-Mart, it could be argued that Wal-Mart is more successful than Carrefour in overall corporate strategy and operation. Carrefour has been shortsighted in its Chinese operation and the success achieved is not able to support its future development in the long run. Carrefour's current CAs are not so strong and actually not sustainable. From its competition with Taiwan-based retailers, it can be found that Carrefour does not have more advantages than them. If it competes directly with foreign-based retailers such as Wal-Mart, it's hard to say if Carrefour will win.

Carrefour has many problems in China. First, its reputation and public image is not good. According to Porter's (1985) classification, Carrefour could be called a "bad" competitor, because it did not obey the national regulations. Although the indirect entry strategy worked initially in Carrefour's favor allowing it to beat Wal-Mart to second-tier cities and to obtain FMA acquiring a large market share in a short time, it ruined Carrefour's brand and reputation seriously. Its management problems, such as poor quality of products and services as well as employees' overtime work, make this worse. Since Carrefour plays in China, it must obey Chinese laws as other retailers do, otherwise, it's hard to succeed in the long term. The second is that its relationships with suppliers have some problems. Building good relationships with suppliers is the key issue for its future success. In the long run, the multi-benefit partnerships with suppliers are necessary. Even China is now a buyer's market and Carrefour has strong bargaining power; using a win-win strategy to deal with the relationships can benefit its CA development and further expansion, and its a building central purchasing system and central logistics system does need cooperation from its suppliers. The third is about the trade-off between its fast expansion and the quality of its development. It is short-sighted for Carrefour just to be concerned with increasing the quantity of stores while ignoring the improvement of its management. If it cannot solve the problem in time, its future CA will be weakened. The forth problem is about central management. Each Carrefour store operates as an independent store, making purchases individually, which may be good in the short-term for fast expansion. However, with the store numbers increasing, it is necessary to have central management. Although Carrefour has done some work on this aspect, the result is still not obvious. Carrefour's future in China largely depends on if it can successfully take central management, particularly in competing with Wal-Mart, which is good at central management. Although Wal-Mart has developed slower than Carrefour in China, Carrefour does not show obvious advantages over it. Another issue is the importance of market share. Although it is important for Carrefour to enter as many cities as it can, it is more important to take as much market share as possible in each city that it has entered. Without enough market share, it cannot enjoy the economies of scale and survive in the intensive competition, particularly in the price war. Its failure in Hong Kong has given the lesson. In retailing, size clearly does matter. Carrefour recognized that only with a big network of stores could it realize sufficient economies of scale to become profitable. The company went into the black only after opening its 20th store. In future, during competition with local retailers, Taiwan-based retailers and other foreign-based retailers, Carrefour may face a serious situation if any of the above problems mentioned above cannot be solved in time.

9.3 Case study: Wal-Mart in China

As early as 1992, Wal-Mart obtained permission from the Chinese central government to develop a retail business in

China; but it did not enter China until 1996, after it finished the complete research on the Chinese market. In October 1996, Wal-Mart entered China and opened a Supercenter and Sam's Club in Shenzhen, a city near Hong Kong.

9.3.1 Wal-Mart's strategy in China

Unlike its internationalization in other countries, Wal-Mart entered China by developing the dual-strategy: the procurement strategy, which is to do procurement in China, making China its important sourcing center; and the development strategy, which is to open stores as it usually does in other countries for internationalization. In another words, Wal-Mart walked into China on two legs: one is to open stores in China; the other is to source products from China, supporting its global outlets, particularly its American stores, the main sources of its corporate profit; and the latter is more important to Wal-Mart in the short term.

The procurement strategy: thinking globally. The procurement strategy is very important for Wal-Mart, because it contributes greatly to its CA of EDLP by taking advantage of rich and cheap Chinese products. Actually, this involves the relationships between its overall corporate strategy and its individual corporate strategy in China. Wal-Mart thinks globally in its Chinese operation. Its Chinese operation must support its overall corporate strategy; it is particularly important for Wal-Mart to keep high growth rate when the American economy is in recession, because about 82 per cent of Wal-Mart's sales and about 75 per cent of its profits are from its American stores. From this point of view, the procurement strategy is a kind of strategic action. Further, the procurement strategy reduces the risk of opening stores in a new market, with different cultures and different consumer behaviors, by making profit from procurement. By 2001, its annual sales in China were just US$400 million; while its procurement amount was over US$10 billion including direct and indirect sourcing, which contributed at least US$1 billion profit; just in Guangdong Province, the procurement was US$8 billion. The procurement not only supports its global operation but also strongly finances its Chinese operation. Its "thinking globally" strategy makes Wal-Mart "kill two birds with one stone."

The development strategy: operating locally. Wal-Mart also pursues the development strategy while it takes the procurement strategy. By June 2003, it had opened 27 supercenters, 4 Sam's Clubs and 2 neighborhood stores in China. The neighborhood store opened in 2001 was the first of this kind of format outside America. Its two strategies co-operate so well that they enable Wal-Mart to achieve the least risk with the largest benefit. Through the procurement strategy, Wal-Mart not only supports its American operation, which is safer and involves less risk than just opening stores, but also makes Wal-Mart understand local markets from its local suppliers. The good relationships developed with suppliers could benefit its further expansion. Meanwhile, by the development strategy, it develops new suppliers in different regions for exploring new products, which could benefit its global procurement. Its experiences obtained from developing the Chinese market could contribute to its further internationalization.

9.3.2 Wal-Mart's entry mode strategy and expansion model

In China, although the competition environment is somewhat flexible, for instance, there is no land planning regulation, no hour limitation in store opening, etc., retailing policy was still strict before December 2001 when China entered the WTO. In the strictly controlled market, the best entry mode strategy normally is to takeover existing retailers. However, by Chinese retail regulation, the only legal choice for Wal-Mart to enter China was by JV. Then Wal-Mart entered China by the mode of JV, which was different from what Carrefour did.

9.3.3 Wal-Mart's development strategy

(1) Target segment in China. Wal-Mart mainly targets the M Class and the O Class; meanwhile, it also targets the small business market for wholesale business. (2) Retail formats. Wal-Mart has three formats in China: Supercenter, Sam's Club and Neighborhood Store, but most of its stores are Supercenters. (3) The transfer of sustainable CAs. In America, Wal-Mart develops its CAs by combining the FSAs and CSAs. However, when it enters China, the CSAs are impossible to transfer for they are location based; while some of its FSAs cannot be transferred either, for the same reason.

Among its FSAs, only its strong brand is non-location bound and is completely transferable. Wal-Mart's corporate culture and management know-how are some location bound FSAs and partly transferable. For example, its important incentive mechanism, the share-purchasing program, cannot be transferred, because the program cannot be implemented in China. Further, Wal-Mart's stores in China are JVs; it is difficult for Wal-Mart to transfer its own corporate culture completely. Its management know-how may not work for the different culture and customer demands. On technology side, the supply chain is a location-bound FSA and hard to transfer in the short term. Its non-location bound FSAs could be transferred across national boundaries for supporting its store development and competition worldwide.

To develop its CAs, Wal-Mart takes advantage of Chinese specific advantages to develop new location bound FSAs. For example, Wal-Mart develops procurement strategy in China. When Wal-Mart takes internationalization, one thing is certain: its EDLP is never changed.

The CEO of Wal-Mart, Mr. Lee Scott, said, "Although people like eating fish in one region and like meat in another, they all like products that are value for money." But Wal-Mart's successful model for developing CA in the USA cannot be transferred completely to China, because the Chinese government defines different competition environments by different regulations. Chinese customer demands are also different from American demands, and both the US-based CSAs and some FSAs could not be transferred to China. It must develop a new model to achieve its EDLP in China. Wal-Mart's CA in China is mainly achieved by developing close relationships with its suppliers, customers, competitors and non-business institutions. Wal-Mart is the center to link them together.

Supplier. There is a characteristic in the relationships between Wal-Mart and its Chinese suppliers: in the USA, it is trading companies that are between Wal-Mart and its suppliers as the intermediates; while in China, Wal-Mart directly faces its over 1,400 suppliers. These suppliers provide Wal-Mart stores with logistics service and JIT delivery. Some of them follow Wal-Mart, going to wherever it sets up stores, with reliable supply systems. This is particularly important when Wal-Mart has not built its own distribution center system in China. This kind of relationships makes Wal-Mart lower its cost and shortens the lead-time. Thus, Wal-Mart reshapes its CA in China. In addition, it also cooperates with Alebaba, a Chinese dot.com company with nearly one million Chinese suppliers, for developing its supplier base. Further, it is developing a new communication system with its local suppliers for creating a culture marked by understanding, commitment and a strong feeling of community to integrate suppliers and improve their efficiency. In China, Wal-Mart just had one warehouse in the Yantian port of Shenzhen city by 2002, which served as the distribution center distributing goods for stores in Guangdong Province. For stores in other provinces, it is its suppliers that undertake all distribution work. Without the support of suppliers, Wal-Mart is unable to develop its EDLP CA in China.

Customer. By providing its customers with value for money commodities and appropriate services, Wal-Mart gradually obtains their loyalty. More important, through developing close relationships with its core customers, the M Class and O Class, and responding to their feedback, it is meeting their demands well.

Competitor. By learning from its competitors, Wal-Mart gradually understands local markets and adapts to local customers' tastes well. For example, it learned from its competitors to sell local taste food and changed the ratio of non-food items to food items from 70/30 to 50/50, which makes its performance greatly improved.

Other companies. Wal-Mart develops strategic partnerships with two companies: Beijing Zhongshanhuihua (BZSHH) and SITIC. The former takes market research for Wal-Mart and promotes Wal-Mart to the local governments before Wal-Mart enters local markets; while the later is in charge of building stores for Wal-Mart and then leasing the stores to Wal-Mart as well as employing employees for Wal-Mart (Figure 8.9). The three companies have very good relationships, such as the current Vice CEO of Wal-Mart

China is the former CEO of SITIC and the Vice General Manager of SITIC was the former Director of Board of BZSHH (Figure 9.3).

Non-business infrastructure. Wal-Mart co-operates with many non-business infrastructures to improve its own image and build good public relationships. For example, it often invites government officers to visit its American headquarters to understand the company, funds schools and is involved in environment protection, etc. It also wins a good reputation by obeying Chinese regulations when developing its Chinese market, which is one of its advantages compared with Carrefour.

Wal-Mart's localization. Wal-Mart has strong advantages in management skills and capital resource. However, these advantages cannot be transferred effectively to the Chinese market until it deeply understands Chinese culture, particularly local consumer behavior. Localization mainly includes the localization of management, the localization of operation and the localization of assortment.

Figure 9.3   Wal-Mart's cooperation oriented expansion model

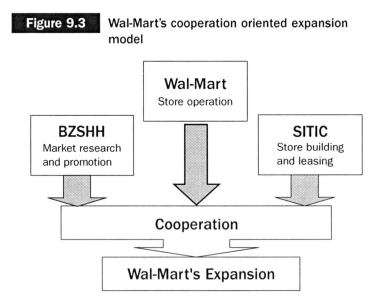

The localization of management: This is the core issue in localization. Since Wal-Mart entered the Chinese market, some progress has been achieved in its localization. For example, Wal-Mart has 6 stores with about 3,000 employees in Shenzhen; most of them are local people. Before the first store was opened, it took over 8 months to train its management team. The company focuses on developing the management team within its employees. It solves the lack of management personnel by changing its initial principle from "obtain, keep, and then develop" to "keep, develop and obtain." In Wal-Mart, all its employees must be trained for at least three months before they start work.

The localization of operation: Wal-Mart tailors its concept to adapt to the Chinese market. Its first Chinese store, Shenzhen supercenter was a milestone in Wal-Mart's history. It was Wal-Mart's first multi-level supercenter in the world situated within a four-tower, 30 storey residential complex to adapt to Chinese conditions. Many changes in its operation have been made to accommodate Chinese customers and different traffic patterns in its stores. For example, unlike Western customers who normally go shopping at the weekend, driving their cars, many Chinese go shopping everyday by bicycle or bus. Chinese customers shop from stores more frequently and purchase smaller quantities. Wal-Mart adapts its shopping bag to fit their needs. In China, the Chinese are very selective with products and demand freshness for food. So Wal-Mart sells live fish, chicken, shrimp, etc; Wal-Mart emphasizes food, and the presentation of food and freshness is superior to its American operation. Food is the most important part of the assortment in its Chinese stores and accounts for about 50 per cent of the sales. In its stores, Wal-Mart adds specialty counts, which is completely different from what it does in America.

The localization of procurement: Local products meet local people's tastes better. Wal-Mart sources over 95 per cent of its products from local suppliers. It has over 1,400 suppliers in China. In recent years, it can be easily found that Wal-Mart's procurement managers are devoting themselves to various kinds of trade fairs in China. In December 2001, Wal-Mart even moved its global procurement center from America to Shenzhen. On the 3rd December 2001, the new global procurement center started to work there. Based on this center, it will build another 20 sub-centers around the world. Within one or two years, Wal-Mart will stop its outsourcing work; all procurement work will be undertaken by the Shenzhen center and other sub-centers. In China, its growth rate of procurement increases by 20 per cent each year, which makes China one of its largest supply bases.

9.3.4 Wal-Mart's growth strategy

In developing the Chinese market, Wal-Mart faced a dilemma if it still continues to keep its golden rule, which is building a distribution center first before opening stores. Without its own distribution centers, efficient operation is hard to achieve, while building distribution centers not only takes much time and huge costs but also has some practical difficulties, such as foreign companies were not allowed to be involved in the logistics business. Although Wal-Mat's successful experience is mainly based on its own distribution system and it has got used to operating with its own distribution centers, it has to operate in an environment without central management and its own distribution system, while asking local suppliers to provide distribution services. This way restricts its expansion speed but is safe and meets Chinese conditions. From its development, it can be found that it is intending to develop the "3 plus 1" model, which is that one Sam's Club is built followed by three supercenters around it in the same city, with the Sam's Club undertaking some of the functions of a distribution center. But it does not give up on building its own distribution system and just waits for the right time. Unlike Carrefour, which quickly entered the richest region, Eastern China, where the competition was more intensive, Wal-Mart liked staying in the second-tier cities where the competition was less intensive. Wal-Mart held its strengths waiting for the further deregulations of the industry. Wal-Mart's growth strategy was more cautious compared with Carrefour; each time, it just entered one city. It did not care about how many stores opened but tried to guarantee that each store opened successfully. After succeeding in one city, then it expanded to another. It was concerned with the quality of the expansion rather than quantity, which explains why it opened only 8 stores, compared with the 27 stores of Carrefour within the first 5 years.

Wal-Mart takes two ways to grow: market penetration and market expansion. It mainly took market penetration at first. Now its pace of market expansion is accelerating.

Market penetration: Wal-Mart's EDLP is a powerful inducement for customers, by which it attracts customers to visit its stores more frequently and buy more. When it enters a city, Wal-Mart tries to dominate the market by taking more market share and opening more stores than its competitors. In Shenzhen, it had opened 9 stores by February 2003. None of the MNRs have opened so many stores in one Chinese city.

Market expansion: Wal-Mart first stayed in Shenzhen; then expanded to the Pear River Delta Area. It is marching to Northern China and will open stores in Beijing in 2003, where Carrefour had opened 4 stores, and other cities, such as Wuhan and Shanghai.

9.3.5 Wal-Mart's problems

It can be argued that the strategic mistake Wal-Mart has made in China is that it changed the target market from Shanghai to Shenzhen when it entered China. It lost the FMA in the richest region in China, the Yangtze River Delta Area centered at Shanghai. In Shanghai, the largest city with a 14 million population, there are over 40 hypermarkets, but there is no Wal-Mart's name among them. Besides this, Wal-Mart has several other problems.

Localization has not been successfully realized yet: Actually, Wal-Mart's most serious problem is still in its localization. Its operation in China indicates that it lacks international experience and has a weak ability in adapting to local markets. Before Wal-Mart entered China, it spent several years doing market research. Yet, when its first store opened, most of the cooked food was still western style, which made Wal-Mart less competitive to local retailers. Then it had to learn and imitate the operation of local retailers to sell local food. "The incident of Wal-Mart's Dongguan store" (Dongguan is a manufacturing center in Guangzhou Province) gives further evidence of its unsuccessful localization. Its Dongguan store rented out some counters to local food producers for selling their cooked food in the Wal-Mart store. However, it was not able to manage this kind of operation well. Some food producers were not even qualified to make food and some cooked food were of bad quality, which caused a lot of customer complaints and Wal-Mart was fined RMB 45,000 on 8th 2002. When the incident happened, Wal-Mart could not deal with the problem in time and properly, which annoyed customers further. The incident indicated that its decision process had problems. During the incident, its frontier employees (not called associates in China) tried to get rid of the responsibility instead of dealing with the problem from the perspective of protecting the company's reputation. This suggested that Wal-Mart's corporate culture had not been transferred to its Chinese operation. Similar incidents of poor product quality also happened in other stores. This indicates Wal-Mart still has a long way to go in localization. In addition, its management personnel also need to be further localized. In Wal-Mart's Chinese companies, the majority of senior managers are American, while middle management teams are normally from Hong Kong. This causes two problems: One is the high management cost. Chinese retailing is becoming a low margin business; its high cost in management teams has been one of the most important reasons for Wal-Mart's loss of money in China. The other is that Wal-Mart cannot realize localization completely in its local operation and it frequently has to imitate local retailers.

The relationships with suppliers have some problems: When it just entered China, Wal-Mart used the buy-out method in its sourcing to achieve low price. This was one of its CAs compared with other retailers whose payment term was normally 60 days. However, since 31st January 2000, Wal-Mart has changed the payment terms and extended them to be as long as 60 days. Meanwhile its requirements for suppliers are becoming much stricter to reduce suppliers profit greatly. In the notice given to its suppliers on 30th January 2000, it not only extended the payment terms but also asked for discount in buying. Also, it asks all suppliers to sign agent contracts making Wal-Mart their agent. Also, Wal-Mart asks that all suppliers must provide products with the lowest price; otherwise Wal-Mart will reconsider their relationships. Wal-Mart also asks all suppliers to support its promotion activities and supply it unique brand products specially made for Wal-Mart. Since 1998, Wal-Mart applies serious punishment to those suppliers who cannot deliver goods in time and use UPC codes correctly. All of these rules benefit Wal-Mart itself greatly while making it hard for many SM sized suppliers to survive. It is not a win-win strategy. In the long run, this may not only weaken Wal-Mart's reputation but also weaken its competency in increasingly intensive competition. A multi-benefit supplier relationship is a potential CA. It can be argued that Wal-Mart is becoming short-sighted.

Too selective goods and assortments: Wal-Mart's products are too selective and its assortments cannot meet the Chinese demand for "one-stop shopping" well. This is one of the most important reasons why Wal-Mart lost to Vanguard in the competition in Shenzhen. Wal-Mart's strategy of 80/20 works well in the States; however, in China, people's buying behavior is different. The Chinese like to fully compare products that they intend to buy before their buying behavior is finished. Their buying behaviors are finished by a full selection, while Wal-Mart's assortment structure is too narrow. Comparing an 18,000 square meter Wal-Mart store with a 15,000 square meter Vanguard's GMS, Wal-Mart just has 15,000–20,000 categories products while Vanguard has 70,000–80,000. Wal-Mart focuses on the sale of a single category so much that it ignores the sales of the total categories.

Concerning price more than quality: Between quality and price, Wal-Mart always tends towards price. It likes to promote its low price image; but low price also means low quality. For example, other retailers sell first class bananas for RMB 8 Yuan/ Kilogram while Wal-Mart sells second-class bananas for RMB 6 Yuan/Kilogram and does not sell the first class ones. Other retailers sell P & G's shampoo, such as Patine brand, which is normally 200 ml per pack, while Wal-Mart sells the 180 ml or 160 ml per pack to a achieve cheaper price.

Depending on IT too much while employees' subjective initiatives are ignored: Wal-Mart is not flexible and presents little adaptability in terms of its marketing mix and management. It asks its employees to follow orders from computers. Because Wal-Mart depends on IT too much while its information system in China is not as complete as that in the States, it cannot respond to the Chinese market change in time. For example, store managers of Vanguard can make any decision in time according to any change in the market, such as reducing prices for certain products. The store managers of Wal-Mart do not have this power and must report to the headquarters office in China before they make any decision. By the time they receive the response from headquarters, the situation in the market has changed. In the States, Wal-Mart's JIT management depends on IT; while in China, it should depend on the store manager more because of the different national conditions and competition environment.

Appropriate service versus satisfactory service: Wal-Mart advertises that it provides satisfactory service. Actually, its principle inside is that it just provides customers with an appropriate service. It believes that increasing service will cause increased costs, which weakens its competency of EDLP. If a customer buys a large product, such as a refrigerator, Wal-Mart provides a transportation service, but it is not free; and the further a journey is, the more it charges. While in Vanguard, this kind of service is free. By comparing with other retailers customers gradually understand whose service is better no matter what they claim for their service.

Logistics system: Wal-Mart has not built a complete distribution system in China and its operations in different cities are separate. Because the Chinese logistics industry is less developed and fragmented, even distribution within 500 kilometers is difficult, so the economies of scale are hard to achieve.

9.3.6 Discussing Wal-Mart's performance in China

Since 2000, the unsatisfactory performances in other international markets and the recession of the American economy have made Wal-Mart accelerate its expansion in China. From June to November 2000, Wal-Mart's senior management team led by its CEO, Mr. H. Lee Scott and the CEO of the international division, Mr. John. B. Menzer, visited China once a month on average. Wal-Mart's future success in China depends on several issues: One is if it can successfully transfer some its FSAs, such as the application of IT and the building of a highly efficient supply chain, to China; the second is if it can take fully advantage of Chinese CSAs, such as the fast growth of Chinese retailing, weak competitors and rich supplies of products; and the third is if it can successfully realize localization adapting to diversified Chinese local markets. Its future success also depends on if it can successfully transfer and build its corporate culture in its Chinese operation. Although John Menzer, the President and CEO of Wal-Mart International Division, said, "The acceptance of Wal-Mart culture is beyond belief," the Dongguan incident indicates that the work is easier to be said than done.

Another important issue for Wal-Mart's future success is if it can involve making the rules of game in Chinese retailing by taking its CAs, such as strong financial resource. Most local retailers are tight in financial status and have to extend payment for their suppliers to over 60 days. If Wal-Mart can use its strong financial resource to shorten the term and make it the rule of the industry, it not only obtains better terms from its suppliers, such as cheaper price, but also forces local retailers to follow the terms, which is very unfavorable for them. Wal-Mart also needs to transfer some of its successful experiences in the States to the Chinese market. Just as it avoided competition with retail giants at its early stage in America, it may be better for Wal-Mart not to involve direct competition too early with other retail giants such as Carrefour. The Chinese market is large and there are many opportunities in the cities where competition is less intensive. Trying to avoid intensive competition with retail giants before it is successfully localized and develops successful model and trying to take as many market shares as it can may benefit its further development in China. During the expansion, its saturation strategy and combination strategy could be effective in China, too. A diversified strategy targeting different segments is necessary for dominating the market. Supercenter could be a main weapon for expansion while the neighborhood format could be a complementary tool. In future, the best way for Wal-Mart to grow in China is by acquisition, but it should be successfully localized first. In 2000, its CEO, Mr. Lee Scott, said in Guangzhou that if Wal-Mart wanted to keep its leading position in retailing, it must succeed in China. However, succeeding in the Chinese market is not easy. By now, Wal-Mart's dual-strategy in China is successful, which trades off its short-term interest and long-term interest well, and contributes its retail sales of RMB 5.85 billion Yuan and procurement US$2billion in 2003. However, its operation in China is just at the beginning and it is still too early to conclude if it has succeeded in China.

9.4 Summary

After China opened its retail market, many global retail giants have entered the market by direct or indirect ways. But few of them have developed successful models. They encounter many difficulties during their development of the Chinese market. Chinese conditions, such as the less developed Chinese logistics industry, the lack of retail management personnel, the diversified local cultures, the bureaucratic way of government administration, the high transaction cost, the diversified Chinese consumer behaviors, etc., make it hard for MNRs to transfer their own operation models developed abroad to China and realize localization in the short term. These conditions also weaken some foreign successful experiences or CAs or make them lose their edge in the Chinese market. The case studies suggest that even the world's largest retailer, Wal-Mart, and the world's most internationalized retailer, Carrefour, also face similar problems: neither of them have developed successful models in China after over 6 years operations and still face many problems in their localizations. Wal-Mart in China has not been the complete and the real Wal-Mart; it will still take MNRs much time to develop their CAs and Chinese operation models, which provide Chinese retailers with some opportunities to catch up.

Chinese retailers' behavior

10.1 A profit pattern for Chinese retailers

10.1.1 The "channel fee" oriented model

The Chinese retail market has shifted from a seller's market to a buyer's market since 1998. The dominant power in the market has shifted from suppliers in the planned economy to retailers in the current transitional economy. How to sell products has become a critical problem for many production companies, because few of them have own sales networks; and they have to depend on several key retailers to sell their products. This provides those retailers with strong bargaining power. Meanwhile, Chinese retailers are very weak in their competency, particularly in developing successful models. Most of them make their strong bargaining power the main means to make profit. Their profits are mainly from the non-operation side, such as "channel fees," and favorable payment terms rather than the management side, such as high efficient supply chain. The most popular Chinese retail model is so called a "channel fee" oriented model.

10.1.2 What is the "channel fee"?

Chinese channel fee is a kind of transaction fee that Chinese suppliers must pay their retailers in order to make their products displayed in the retailers' stores for selling. It is similar as slotting fee or slotting allowance existing in the Western countries (In the UK, it is also called "key fee"), which is paid by a vendor for its space in retail stores; but different from it in both the amount and the kind of fee. The Chinese channel fee is not just one kind of fee but a group of fees including many different fees and varies from one retailer to another depending on a supplier's bargaining power; so does the amount of the fee. Charging channel fees has become the rule of the game in the Chinese retailing and the main source of Chinese retailers' profits. Nearly all retailers in China charge their suppliers channel fees. Channel fees have widely existed in Chinese retailing although have long been ignored for neither retailers nor suppliers like discussing them.

10.1.3 How much is the channel fee?

The Chinese channel fee normally includes store entry fee, promotion fee, advertisement fee, exhibition fee, and so on. Some channel fees are often not listed in the contract between retailers and their suppliers and are often charged by retailers without notice. The follwing shows some channel fees charged in Guangzhou market; and some channel fees paid by a large Chinese dairy company, company Y.

Some items of channel fee the retailers charged in Guangzhou City

- *Entry fee*: From RMB 4,000 Yuan to several ten thousands Yuan. All suppliers must pay it.
- *Shelf display fee for new products*: RMB 500 Yuan per product per store. Suppliers pay it when their new products are displayed on their retailers' shelves.

- *Sponsorship fee for new store opening*: RMB 1,000 Yuan per store. Retailers charge suppliers this fee when they open new stores.

- *Store's anniversary celebration fee*: RMB 1,000 Yuan per store. Each year, retailers charge suppliers this fee in their stores' anniversary period.

- *Advertisement fee*: RMB 1000 Yuan per store. Retailers ask suppliers to share their advertisement fees.

- *Ground displayed fee*: RMB 1,000 Yuan. Retailer charges this fee when a supplier displays its products on the ground in the retailer's store.

- *Short-term promotion fee*: RMB 500 Yuan. Retailers charge this fee when they take a short-term promotion.

- *Festival fee*: Retailers charge it in kinds of festivals, such as RMB 1,000 Yuan per store for Chinese New Year; RMB 2,000 Yuan per store for the National Day; RMB 500 Yuan per store for the First May Festival; RMB 500 Yuan per store for the Middle Autumn Festival; RMB 500 Yuan per store for New Year; RMB 500 Yuan per store for other festivals.

- *Commission fee*: 3 per cent of the settled amount of each month. Suppliers must pay this fee each month when they settle payment with their retailers.

- *Discount for new store opening*: 3 per cent of the settled amount at the month new store opened (suppliers must bear loss caused by low price promotion).

- *Compensation fee for product depreciation*: 0.5 per cent of the amount settled each month.

(Note: Y is the second largest dairy company in China and has very strong bargaining power).

For example, the contract between retailer T, a large retailer in the Southern China, and one of its suppliers shows that the supplier's channel fee in 2001 included 30 items of fees, such as unconditional commission fee, conditional commission fee, new year promotion fee, Chinese New Year promotion fee, the Labour Festival promotion fee, Duanwu Festival promotion fee, Middle Autumn Festival promotion fee, National Day promotion fee, the minimum display fee for new products, the minimum fee for sponsoring new store opening, the minimum discount for new opened stores, the minimum promotion fee for store's anniversary celebration, the minimum discount for anniversary celebration, the compensation for product depreciation, the lowest discount for promotion, the minimum promotion times in a year, compensation for new opened store, the minimum promotion fee for shelves, the minimum compensation fee for advertisement, price protection fee, the fee for dealing with returned products that are not sold well in 30 days, the discount for big purchase, and so on. Among these fees, unconditional commission fee was 3 per cent of the total sales, the promotion fee for Chinese New Year was RMB 1000 Yuan per store, National Day promotion fee was RMB 600 Yuan per store, the minimum displayed fee for new product RMB was 500 Yuan per product per store, the minimum discount for new opened store was 3 per cent, compensation for depreciation was 0.5 per cent, the minimum counter fee RMB was 500 Yuan each time per store.

Another supplier's contract in 2001 with retailer T was different, in which the items of channel fee included award for sales, procurement commission, advertisement fee, discount price for new opened stores, compensation for product depreciation, and so on. Among those fees, the supplier agreed to pay the retailer RMB 1000 Yuan entry advertisement fee and RMB 1000 Yuan media advertisement fee for each store; festival advertisement fees are: RMB 1000 Yuan for Chinese New year; RMB 500 Yuan per store for the May Festival, Duanwu Festival, and Middle Autumn Festival; RMB 200 Yuan per store for National Day; Advertisement fee for new product RMB 300 Yuan per product per store; RMB 1000 Yuan advertisement fee for a new store and RMB 1000 Yuan for anniversary; 0.5 per cent of the sales as compensation for the product depreciation; new supplier should pay RMB 1500 Yuan for its entry fee. The contract also asked the supplier to take three times promotion in each store and each time was no less 10 days.

In 1998, Carrefour charged its suppliers at least 14 channel fee items. For example, in the contract between Carrefour and one of its suppliers, the festival promotion fees were RMB 1,000 Yuan per store for the new year, Chinese New Year and the National Day respectively; RMB 10,000 Yuan for its new opened store; RMB 3,000 Yuan per store for the store's anniversary celebration each year; RMB 2,000 Yuan per brand for new product display; 20 days' promotion with free products and high discount (not specified); and there were also other sponsorship fees and the fee for redecorating its store.

The amount of channel fee charged by retailers is surprising. Shanghai Guanshengyuan Group, a food supplier in Shanghai, paid its channel fee of RMB 4 million Yuan in 2000, RMB 4.80 million Yuan in 2001 and RMB 6 million Yuan in 2002 respectively. In Shenzhen, a supplier complained that in 1999, it supplied its retailer RMB 1.62 million Yuan products while just was paid for RMB 1.29 million Yuan after the channel fees charged. In 2001, it supplied the retailer RMB 1.21 million Yuan while was charged for channel fees of RMB 260, 000 Yuan, over 21 per cent of the product value (this is very considerable to retailers comparing just 2 to 3 per cent of net profit margin in Chinese retailing). According to the retailer, suppliers' gross margin normally is about 20 per cent including 16 per cent of operation cost. So the net profit margin is normally just about 3 to 4 per cent. However, after retailers charge over 20 per cent of the channel fees, suppliers are hard to make profit and actually often lost money. Channel fees have become retailers' the most important source of profit in China. For example, a large retailer in Shenzhen with 2,000 suppliers charges at least about RMB 200 million Yuan channel fee each year. Further, when retailers charge suppliers channel fees, they normally give the suppliers informal invoices, which means that the retailers do not pay tax for these revenues. In 2001, the total profits of the top 500 Chinese retailers were RMB 600 million Yuan from their store operation; while their profits from the channel fee were RMB 3,600 million Yuan (*The Southern Metropolis*, 17th September 2002).

10.1.4 How is the "channel fee" generated"?

Channel fee was first introduced in China as an "international rule" by Carrefour when it entered China. It became popular quickly in China and has been widely accepted by Chinese retailers. Actually the real reason is resulted from the current situation of the transitional economy: Firstly, from the market side, the Chinese market has been an oversupplied market since 1998, particularly for basic commodities. Secondly, although retailers have controlled the end of distribution channel in retail process, intensive competition in recent years has squashed the retailers' profit margin to very low level; the competition forces them to grow up as quickly as possible; while nearly all Chinese retailers face a critical problem: the lack of enough capital for growth; thus charging channel fees becomes a fast and easy way to solve the problem. Thirdly, most of suppliers are SM; their bargaining power is very weak. The planned economy made few involved in distribution field and few of them have their own complete distribution system. Suppliers are very fragmented and supplier association has not been formed. Meanwhile, they still focus on production business while do not have experience in retailing or building sales networks. They have not or are not able to involve retail business or developed their own retail outlets. So they have to depend on retailers to sell their products. During the transitional economy, Chinese retailing is also in transition. The government does not have experience in supervising retail competition and there are few regulations or laws regulating the competition. Retail outlets are scarce, hence retailers are free to charge a channel fee because they control the final stage of the retail channel, which is the only way the product can reach the consumer.

10.1.5 The negative influences of "channel fee"

A reasonable channel fee is acceptable. However, the overcharged channel fee brings serious negative impacts on developing the Chinese retailing:

- It encourages unfair competition. Many retailers depend so much on channel fees that they lose motivation in improving retail management, such as in innovation, cost control and supply chain management. Depending on channel fee makes many retailers throw away the basic retail profit model and results in the deteriorating performance of the whole industry. To many retailers, the channel fee even account for 80 per cent of their profits.

One reason that many retailers do not develop their distribution centers is that their suppliers undertake the logistics work transporting goods directly to the stores; while they still charge suppliers' a "logistics fee." According to Ms. Xu Lingling, the financial director of Lianhua, the largest retailer in China, in current Chinese retailing, a retailer's profit comes from five sources: operation revenue from its stores' operation, the profit from the operation of logistics and distribution between its suppliers and the retailer's stores, channel fees, franchise fee for franchising operation, and profit from the discount in procurement. Given the low net profit margin of current Chinese retailing, most of retailers can just achieve break even or a few profits from their store operation; and charging channel fees has become their main profit source. The more stores a retailer opens, the more channel fee the retailer can charge. This encourages the retailer to open more stores blindly without considering its capability such as management and financial conditions.

- Suppliers' interests are seriously jeopardized. Retailers transfer their risks to their suppliers while the suppliers' interests cannot be guaranteed. Channel fees cause unfair retail trade. Retailers use this way to transfer their costs and risks to their suppliers. However, they do not give any promise for the suppliers such as sale volumes; and they return all unsold products to the suppliers. Further, retailers overuse their suppliers' capital; they normally pay their suppliers in 60 days; some even longer. Retailers squash their suppliers' normal profit margin and overuse the suppliers' capital. The more a supplier supplies its retailer products, the more the retailer uses the supplier's capital; many retailers use this capital to open new stores or invest in other businesses, such as in stock market. This pushes the suppliers at very unfavorable financial status. Under this kind relationship, if a retailer goes bankrupt, all its suppliers will be involved and many small and medium-sized suppliers will also go bankrupt for they cannot take their money back from the retailer. The whole supply chain will cause a domino effect and it is a lost-lost result. In 2002, many of this kind of case happened in the Chinese retailing involved the largest retailer in the North-eastern China and one of the largest retailers in the Southern China. Some retailers use opening stores as a way to "borrow" money freely from their suppliers, then invest it in stock market.

- Several large suppliers could control the retail outlets in a region by their strong financial strengths and by squashing SM suppliers out of these outlets, which can bring homogeneous phenomenon in the store assortments and easily causes price war to squash profit margin among retailers.

- Channel fees weaken the function of the procurement department of a retailer, which is the root of retailing, because the retailer likes "purchasing" products from those suppliers who can afford channel fee. Once this function is weakened, the development of whole retailing will be influenced.

- Consumer interest is jeopardized finally, because retailers choose suppliers mainly by channel fee rather than their products supplied. Homogeneous stores could make consumers face less selection and poor quality of products.

- The over-charged channel fee weakens suppliers' development potential. Some suppliers face financial problems caused by the channel fee and have to reduce their product cost by lowering the product quality to save money. The poor product quality makes consumer go away, which often causes the supplier to low prices again for they can only compete for price. Thus a vicious circle happens, which brings negative effects to both consumers and suppliers themselves.

Actually some MNRs also charge channel fee in other developing countries. In June 2002, Thailand government asked Carrefour, Tesco and Wackro to explain the reason for entry fee they charged from suppliers and planned to make regulation to restrict the charge of the fee. In China, because there is no regulation regulating "channel fee," retailers can do what they want; thus the problem of channel fee is more serious than those MNRs charged in other countries.

10.1.6 The unfair retail trade to suppliers

Besides channel fee, there are some other unfavorable terms for suppliers. The most complained one by suppliers is the terms of payment. The most popular payment terms is 60 days, which means retailers pay their suppliers in 60 days later after they receive the suppliers' goods; some are in 90 days. However, even though the time is so long, many retailers still do not pay their suppliers in time, which is a very popular phenomenon in the Chinese retailing. Some retailers even delay to 180 days and use the money to open new stores or invest in other businesses. In recent years, many payment disputes have happened between retailers and their suppliers in China. Many large retailers, such as Vanguard and Wackro have been sued. Retailers normally bear less risks than their suppliers: they charge much channel fee from their suppliers, such as 20 per cent of the value of the goods supplied; they obtain free goods from the suppliers within payment period, such as within 60 days; they return all unsold goods back to suppliers without any payment; and the more goods the suppliers send to them, the more money they "borrow" from their suppliers. Retailers even ask their suppliers to pay taxes for them by charging a channel fee.

10.2 Case study: a failed but popular retail model in China

In China, most retailers' business model is the channel fee based model. Actually it is just a simple retail trader model: a retailer rents places to open stores; all commodities that it sells are "borrowed" from its suppliers; it pays its suppliers many days later after the commodities are sold out and often invests the suppliers' payments in other businesses. When a retailer depends on the channel fee greatly while ignores corporate management and invests its suppliers' capital in other businesses, such as in stock market, the retailer may face a very serious situation once the company encounters financial problems. The case of Huarong Supermarket Group gives an example.

Huarong Supermarket Group was established in October 1995 in Fujian Province. It was once the largest supermarket chain in the Province. It claimed to open 500 stores with over 300,000 square meters before 2003. Because it made the channel fee to be its main source of profit while ignored improving management and developing CAs, its financial status became worse quickly in the intensive competition. The worst thing is that it used its suppliers' payments to invest in other fields but lost most of the money. Then, the financial crisis emerged at the end of 2001 when it could not pay its suppliers payments on time. Over 100 suppliers came to the company asking for their payments that had been delayed for over 3 months, some even for 8 months. At 9:00 pm on 27th December 2001, a crisis finally happened.

Some angry suppliers wanted to go to the supermarkets and take their goods back. When the news was spread, more suppliers went to the company asking for their payments; and over 400 suppliers stopped supplying the company. Meanwhile customers, who had the company's IC card (customer put their money in the card first, then debited from the card when they bought product from the company), crowded to the supermarket to try spending all the money that they had put in the IC card. Suddenly, most of the products on shelves were bought. Many customers had money in their cards that they could not spend as nothing was left on the shelves and all the suppliers had stopped supplying the company goods. To solve the problem, on 7th January 2002, the company held a meeting with its suppliers and proposed either the suppliers become the company's shareholders and operate the business with it together or the company went into liquidation to pay them. Actually, the company had owed its suppliers RMB 55 million Yuan while its total assets were just about RMB 30 million Yuan. Since it had lost its credit, none of its suppliers liked to co-operate with it. Therefore, it had no choice but bankruptcy, which also made some of its suppliers bankrupt. Huarong became the fourth largest retailer bankrupted in the area for payment reasons. The similar cases also happened frequently in other regions such as Beijing, Shanxi Province, Zhejiang Province.

10.3 Chinese retailers' growth model

In recent years, Chinese retailers have developed very quickly. They mainly take four paths for their expansion: organic growth, growth by franchising, M&A or JV. For example, Lianhua takes all of the four ways in its expansion (see the case of Lianhua). But most Chinese retailers make organic growth the main way for their growth, because they have not developed significant brand images, standardized operation models and enough management experience. Some large Chinese retailers with successful models and core competencies try using M &A for fast growth. However, they often face two problems: One is the financial problem. They do not have enough capital to fund their growth, particularly for M&A, due to their past history. In the past, the central government was mainly concerned with investing in the production field while it ignored investing in retailing; and the profits made by retailers were all handed over to the government. Thus retailers have limited capital for further development. Retailers normally have three ways to solve the financial problem: either to obtain loans from banks, or build JVs with foreign investors, or to be listed on the stock markets. Being listed on stock markets has become popular in recent years, because this way normally provides more money than the other two ways with longer-term use of the money. But only a few Chinese retailers are qualified to be floated. Even the largest retailer, Lianhua, was not listed until 2002. The other problem is that when they take M&A, they often fall into the trap called diseconomies of scale in their M&A: They cannot achieve economies of scale after the M&A; in another words, after the M&A, they cannot achieve significant cost reduction. A retailer's cost mainly includes four parts: procurement cost, logistics cost, operation cost and rent cost. Among these costs, procurement cost normally accounts for about 80–85 per cent, logistics cost accounts for 4–6 per cent, operational cost accounts for 4 per cent and rent cost accounts for 3 per cent. This means procurement and logistics account for about 90 per cent of the total cost. Therefore, only when the two companies' procurement systems and logistics systems are integrated, can they benefit from their M&A; or it can be said that if the retailer cannot successfully integrate the two companies' procurement and logistics systems, the M&A would not succeed. In China, the supply chain is very local and fragmented, there are few partnerships built between suppliers and retailers, and many Chinese retailers have not developed a successful management model; therefore, expansion by M&A is full of risk.

Thus, most Chinese retailers think the best way for growth is organic growth rather than M&A.

During their expansion, it is important for Chinese retailers to achieve the balance in both IF and OF. IF concentrates on suppliers and internal operation while OF concentrates on customers and direct competitors. We will now examine three different case studies to illustrate the different developing paths of Chinese retailers including their IF and OF.

10.4 Case study: Lianhua Supermarket Corporation Limited

When we talk about Chinese retailing, we have to mention Lianhua, the largest retailer in China. Lianhua Supermarket Co., Ltd. was established in May 1991 as a government project by the Shanghai Commercial Commission. It was named Shanghai Lianhua Supermarket Commercial Company with the investment of RMB 1.2 million Yuan. Its first supermarket was opened on 21st September 1991, and was just 800 square meters in size with 2,000 categories of commodities. The supermarket actually was copied from Wellcome, Park' N Shop and China Resources Store, the three Hong Kong supermarket chains. Before that, nearly all supermarkets in Shanghai were JVs with Hong Kong investors and targeted upper-class segments, while Lianhua targets the mass market. The supermarket attracted so many customers in the first month that they had to line up to enter the store. The success of the first supermarket greatly encouraged the company. Then Lianhua took chain operation for development. Since then, Lianhua has developed at an amazing speed; its store number increased from 1 in 1991 to 2505 in 2003; and annual sales increased from RMB 200 million in 1995 to RMB 24,031 million in 2003. Its net asset had increased by over forty times from RMB 12 million Yuan to over RMB 500 million Yuan in ten years. Its sales in 2003 reached RMB 24 billion Yuan.

In 1995, Lianhua built a JV, Shanghai Carhua Supermarket Limited, with Carrefour to open hypermarkets in Shanghai, in which Lianhua took 45 per cent of the shares. In January 1996, the first JV, Carhua hypermarket was opened. Lianhua first introduced Carrefour's concept of the hypermarket in Shanghai. It was the largest store in Shanghai with over 40,000 square meters. On its first day, over 50,000 customers visited the hypermarket and the sales were over RMB 700,000 Yuan. By 2001, there had been six Carhua stores opened in Shanghai. By 2002, Lianhua had over 1,300 stores with annual sales of RMB 14 billion Yuan. On 24th March 2000, the 1000th chain store of Lianhua supermarket was opened. Since 1997, it has been the leader in Chinese chain store operation. In 1999, Lianhua replaced Shanghai No.1 Department Store with annual sales of RMB 7.3 billion Yuan to be the largest retailer in China. Shanghai No.1 Department Store had been the largest Chinese retailer with one independent department store for decades. In the Chinese retailing, it was in first time that a supermarket chain replaced a department store to be the first place. This indicates that the dominant position of Chinese department store has been weakened and the supermarket times are coming in China. From then on, Chinese retailing has entered the diversification times in retail formats. To attract more investments for further development, Lianhua reformed its ownership structure from a 100 per cent state-owned company to a limited company in 1997. At the end of 1998, it changed its name to Lianhua Supermarket Ltd.

In 2000, the main shareholder of Lianhuan, Shanghai Friendship Group Corporation Limited, which is a listed company, restructured the company. The current ownership structure is shown in the following:

Lianhua's development can be divided into three stages:

1. Start up stage (1991–1995): The establishment of Lianhua was supported by the Shanghai government. The

Table 10.1 The ownership structure of Lianhua in 1997

Shareholders	Stake
Shanghai Friendship Group Co., Ltd.	40 per cent
Shanghai Industrial United Co., Ltd.	30 per cent
Mitsubishi Co.,	15 per cent
Wong Sun Hing Investment Ltd	4 per cent
Shanghai United Trade Co.,	5 per cent
Lianhua Employees	6 per cent

Table 10.2 The ownership structure of Lianhua in 2002

Shareholders	Stake
Shanghai Friendship Group Co,. Ltd.	51 per cent
Shanghai Industrial United Co,. Ltd.	31.73 per cent
Mitsubishi Co,.	10.10 per cent
Wong Sun Hing Investment Ltd	4.23 per cent
Shanghai friendship-Fortune Corporation Ltd.	2.94 per cent

capital needed for its development was loaned by banks while the government paid for the interest. It targeted the Shanghai urban market and had lost RMB 5 million Yuan by 1995. During this period, its main goal was to explore a successful retail strategy.

2. Fast growth stage (1996–2000): Growing to be the largest retailer in China. In 1997, it built 14 Lianhua JVs with other Chinese retailers. The growth model was alliance strategy. It focused on suburban and neighboring regions. Since 1996, it started to make a profit when the store number reached 108. By 1999, the profit reached RMB 53 million Yuan with RMB 7.3 billion Yuan sales. In 2000, it built a JV with the largest retailer of Zhejiang Province for opening stores there. By July 2000, its JVs reached 23 with 215 stores, over 40 per cent of the total stores.

3. Innovation stage (2001–2003): With China entering the WTO, Lianhua starts to march out to the Yangtze River Delta and takes a national strategy, focusing on the whole nation. It needs a new strategy for the WTO times.

10.4.1 Lianhua's corporate strategy

Target market: Lianhua mainly targets the O Class, the value oriented segment, for their basic consumption. This is the largest consumption group in China. Lianhua meets their demands by providing them wide selections of merchandise at reasonable prices, particularly by its PL products. Besides this, it also targets the M Class.

Retail format. Lianhua uses hypermarkets, supermarkets and convenience stores to reach its target segments. It planned to launch discount stores in 2003.

Sustainable CA. Lianhua develops its sustainable differentiation advantage and cost advantage in its unique ways:

Differentiation advantage. In recent years, a noticeable phenomenon in Chinese retailing is the homogeneous commodities among Chinese retailers. The board director of Lianhua, Mr. Wang Zongnan, said in 1999 that the differentiation rate of products in Shanghai's supermarket was just 3 per cent, which indicated that all supermarkets had 97 per cent of commodities in common. People often say, "thousands stores are with one face": Similar product assortments, similar prices and similar services. Under that situation, if any retailer could successfully develop differentiation CA, it would attract more customers and take more market share than others. So Lianhua develops its differentiation CA by its food-focus strategy, PL products strategy and service-focus strategy. Its principle is to create deeper and wider assortments than its competitors: "I have the products that you do not have; if you have the products, then I also have them but either my products are cheaper or better than yours or we provide more product selections than yours," said the General Manager, Mr. Liang Wei.

(1) *Food focus strategy*: Concentrating on developing food departments, particularly fresh food departments is Lianhua's sharp competitive edge. Food sales account for over 30 per cent of its total sales, the highest rate among Chinese retailers. In December 1995, in order to support the local government's "Vegetable Basket Projects," whose goal was to provide local residents with a rich selection of fresh vegetables (Shanghai residents traditionally faced a shortage of supplies in fresh vegetables), it launched the project called "Vegetable Basket in Supermarket" and "Ketch Project" providing customers with fresh vegetables and food. It first introduced fresh food into supermarkets in Shanghai, which just met the consumer demand for fresh food, while other retailers did not do this because of the high risk, high cost, low profit margin and lack of experience in the operation.

Developing fresh food not only attracted more customers for shopping but also made the total sales nearly double in each store. In March 1996, Lianhua built fresh food departments particularly focusing on developing fresh food. It was a strategic decision and fresh food has become its main CA. In May 1998, it opened the first fresh food supermarket and achieved great success. By early 2000, there were 350 stores operating fresh foods and the total gross profit was RMB 220 million Yuan. Fresh food reached 1,100 categories included over 500 PL products. In its fresh food supermarket, the food structure has changed from fresh food to normal food to manufactured products from 3:3:3 to 7:2:1. It will open 300 fresh food supermarkets in Shanghai by 2003.

To develop product assortment, Lianhua established nationwide sourcing networks. For example, it built procurement centers in Jilin Province, Zhejiang Province, Guangdong Province, Yunnan Province, Sichuan Province and Shandong Province; it has introduced over 4,000 local specialities in its stores. Meanwhile, it sources special locally famous products across the country. Its typical 10,000 square meter supermarket normally provides about 20,000 categories of products selected from about 80,000 categories of local products. Actually, few Chinese retailers have nationwide procurement networks. It performs the assortment function better than other retailers by sourcing more special local products than its competitors. By sourcing locally famous foods from different cities, its stores have become the food collection center providing wide assortments for customers' selection. It develops the strategy called "Two Fresh plus Two New." "Two fresh" refers to fresh vegetables and fresh health care products, while "Two New" refers to new products imported from abroad and new prices by re-pricing.

To develop the fresh food operation, Lianhua cooperates with its suppliers to build national source networks supplying famous local products. For example in January 2000, it sourced about 300 categories of products from Yunnan Province; in May it imported 18 of categories of fruit from Thailand. In September, it introduced 500 categories of green foods from Jilin Province. In this way, it improved its differentiation rate to 10 per cent and non-local products to 30 per cent compared with other retailers. It took public bidding nationwide for its fresh foods procurement. For example, in 2000 a company from Shandong Province won the bid for supplying 2,000 tons of superior quality apples by beating 24 other companies. It also holds various exhibitions in its stores to promote new products. For instance, it held "Southern China Food Exhibition Fair" and introduced over 500 Southern specialties. Those fairs achieved great success. The change rate of its products keeps 30 per cent each year. Besides regular promotions for certain products, Lianhua held "The First Fresh Food Festival" in its hundreds of stores from December 27th 1997 to January 10th 1998. In 1999, it first held the "National Famous Local Foods Festival" in its stores. These festivals have become its regular promotion activities. In addition, in order to meet consumer demand better, its assortments vary in different store sites. In a resident district, its stores focus on fresh food, while in the city center, the stores have more imported products targeting the M Class. At the same time, it keeps changing product assortments. From 1991 to 1998, it eliminated over 3,000 categories of unpopular products and introduced over 5,000 new categories.

(2) *PL product strategy*: Lianhua greatly develops its PL products, which are up to 30 per cent cheaper than similar products. Lianhua develops two brands of PL products covering fresh food and small general merchandise: "Lianhua" brand and "Chunyi" brand. Just in 1999, it developed 184 categories of PL products and the total sales

reached RMB 220 million Yuan. Some of them have become the most popular products in Shanghai. Each year it eliminates 10 per cent of the PL products and replaces them with new products so that the products can meet customer demands well. Lianhua had over 1,200 categories of PL products by 2001 and the number keeps increasing by 30 per cent each year. These products boost more sales (40 per cent more) with higher profit margin (8 per cent higher) for its lower price (averagely 17 per cent cheaper). Meanwhile, it builds customer loyalty and good brand image and develops OF.

(3) Service focus strategy: Lianhua sets the corporate slogans: "for the customer, customer convenience and customer benefit" and "Lianhua is at your side," which are reflected in its service. For example, as early as 1996 it had developed 45 kinds of services including clothes washing, film printing, free tea, free umbrella, electrical appliances repair, etc.

In addition, to develop differentiation advantages, Lianhua keeps learning. In 1996, when the former General Manager, Mr. Wang Zongnan, visited Carrefour in France, he found about 25 per cent of products sold there were PL products. When he came back, Lianhua started launching its PL project. He also learned to print produce date on eggs from Carrefour, which is popular in western countries but is not in China. Then Lianhua became the first Chinese retailer to sells eggs with produce dates. Meanwhile, Lianhua emphasizes innovation to improve its operation. In 1998, Lianhua realized three "firsts" in Chinese retailing: first to obtain ISO9002 international quality qualification; first to build the most advanced and largest distribution center; and first to establish MIS. Innovation has been one of its main advantages.

Cost advantage. Supported by the three "networks": national sales networks, national procurement networks and national MIS, Lianhua develops the cost CA from three sources: procurement, management and distribution.

(1) *Procurement*: Advanced purchase technology and nationwide purchasing networks make it realize economies of scale. It builds many supply bases in different cities, takes central purchasing and sources products directly from suppliers in the country. In the supply relationships, it tried a new contract model and made a new policy, such as no entry fee in the first three months, buy out 1,000 categories of FMCG without merchandise returned. In this way, Lianhua achieves low procurement price. In Chinese retailing, most retailers pay suppliers in 60 days, some even in 90 days. When Lianhua takes the buying out, it pays those suppliers directly the procurement finishes and thus it can obtain favorable prices from the suppliers.

(2) *Distribution system*: In China, the logistics industry is very undeveloped, but Lianhua has its own advanced distribution systems, especially distribution centers. It built the first IT based intelligent commodity logistics center in China. At the end of 1998, its largest distribution center in China, a 35,500 square meter informationized distribution center, started to work, which is able to supply 30 stores simultaneously with 6,000 cantons of products and be finished within 40 minutes in Shanghai. This efficient distribution system brings Lianhua great CA.

(3) *Management*: During its operation, Lianhua keeps improving its management to develop low cost operation. In August 1998, it obtained the ISO9002 certificate and became the first Chinese retailer that won the qualification, by which it standardizes its operation and quality management system. In November 1998, it first built MIS in Chinese retailing. Lianhua's low cost is also achieved by economies of scale from its multiple retail formats operation and fast expansion.

Lianhua's development strategy. Lianhua's development strategy is simple: speed plus scale. Lianhua's theory is that "the failure of some MNRs in China is due to their low expansion speed," Mr. Wang Zongnan said. So it takes a fast growing strategy. In its development, Lianhua faces a problem: stay in the city center, or go to the suburban areas where most residents live. Lianhua chooses the latter. At the early stage, Lianhua was weak and growing stronger was important. So Lianhua took the strategy of using the countryside to surround the city. Developing in suburban areas is cheaper because of lower property price and less competitive as there are fewer retailers.

Fast growing strategy Lianhua's first strategy is to grow as quickly as possible so that it becomes stronger to compete with others. From 1993 to 1996, Lianhua opened a store every 15 days. It has two growing models: alliance model and acquisition model.

(1) *Alliance model*: This model is to ally with other partners to realize speedy expansion. Lianhua takes different investment methods by different formats. It mainly takes direct investment for supermarkets, franchising for convenience stores and JV for hypermarkets. This makes Lianhua take full advantage of all kinds of resources. By building JV with Carrefour, Lianhua learns how to operate hypermarkets; by building JV with local partners, it solves the problem of localization. Besides MNRs, Chinese retailers also have the problem of localization when they expand into a new region. Because Lianhua's own resource is limited, alliance allows the company to take advantage of more resources from other companies for its fast growth. The nature of the alliance model is that Lianhua exports its brand and management and uses these intangible assets to exchange tangible assets with other companies such as real estate. Building JV is Lianhua's powerful weapon beating competitors. Over 40 per cent of its

stores are built by JVs. The cooperation with Carrefour makes Lianhua learn experiences in operating hypermarkets; then it opens it own hypermarkets outside Shanghai where Carrefour cannot operate because of policy restrictions.

(2) *Acquisition model* Lianhua also takes acquisition for expansion. This is mainly in Shanghai, because good sites for store opening are rare in Shanghai, or they cannot be found in some areas. By takeover of existing stores, Lianhua obtains sites where it wants them. In addition, by acquisitions, it obtains logistics companies or a distribution system to support its chain operation. In 1999, it acquired over 100 stores and distribution centers.

Lianhua's strategies supporting fast growth. To grow fast, Lianhua needs enough finance, management teams and technology. It achieves them by developing capitalization strategy, human resources development strategy, technology leading strategy as well as branding and personalization strategy (Figure 10.1).

Lianhua's diversification strategy. This includes retail format diversification and business diversification: *Retail format diversification.* It diversifies its retail formats to meet the customer demands of different segments. Lianhua operates four retail formats:

- *Hypermarket*: Over 10,000 square meters in size
- *Reinforced Food Supermarket*: 3,000–5,000 square meters
- *Standard Supermarket*: 600–1,200 square meters
- *Convenience Store*: About 100 square meters

Among them, its convenience store is the fastest growing because this format is just at the starting stage in China. It is about to launch discount store in 2003. Hypermarkets, supermarkets, convenience stores and the new discount

Figure 10.1 Lianhua's strategies for fast growth

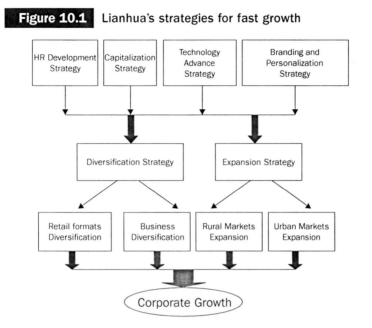

stores enable Lianhua to adapt to various opportunities in Chinese market.

Business diversification. Lianhua also diversifies its business and enters new fields, such as e-commerce and the pharmaceutical industry. On 26th January 2000, it built a JV with Shanghai Fuxing Pharmaceutical Ltd for developing a drug store chain. In 2000, it entered e-commerce business and planned invest RMB 500 million Yuan to be an e-tailer. In September 2000, Lianhua built the JV called Shanghai Lianhua Electronic Commerce Ltd with other companies: www.lhok.com, the largest supermarket website by supplying over 3,000 categories of products by the end of 2000. Consumers can shop on the Internet or dial 96801 for delivery and shopping.

Market expansion strategy. Centered at Shanghai, Lianhua first expanded in the Yangtze Delta Area, then into the neighboring provinces. Now it marches to Southern

China, such as Guangdong Province. It also targets rural markets near Shanghai and opens stores in the suburban areas that are less competitive and ignored by other retailers. Meanwhile, it also takes market penetration. By holding exhibitions of local famous brand products sourced nationwide and by stressing the development of fresh foods, it attracts customers to visit its stores frequently. Expansion needs capital and human resources. Lianhua takes human resource development strategy to meet the personnel demand of fast expansion, which includes building a training school for employee training, cooperating with universities for manager training and developing a promotion system to promote management team within. It also takes capitalization strategy to finance its development, which includes corporate restructuring and M&A. For example, in 1997, Lianhua restructured to become a limited corporation, by which it obtained over US$10 million international capital supporting its expansion. Then it was listed on stock markets in 2002. To manage the fast growing business and improve efficiency, it develops technology-leading strategy by wide application of IT, such as MIS systems in management and distribution. Lianhua also develops personalization and branding strategy to win customers. By focusing on fresh foods and developing PL products, it builds good brand image and customer loyalty as well as improving its profit margin.

Lianhua is very ambitious to grow. It is stepping out of the Yangtze Delta Area and expanding to the whole country.

10.4.2 Lianhua's problems

Lianhua often claimed that the lack of capital greatly restricted its further expansion. Since 27th June 2003, when Lianhua was listed on Hong Kong Stock Exchange, the problem of lacking of capital may not be the most important. It may have at least three main problems. The first is the lack of enough talents. This may be a main reason why it takes JVs for expansion. Enough qualified talent is the most powerful CA. Faster expansion means more qualified management teams are needed. Lianhua could open six stores in one day, but cannot produce a professional manager in one day. Without enough qualified talent, all CAs lose their roots. The second is how to deal with the relationships among management, speed and scale. When more stores are opened, how to keep management efficient or enhanced is important. It may be more important to guarantee that each store opened is successful rather than just more stores are opened in its expansion. The third is in its diversification of formats. Lianhua tries to cover all retail formats. It is doubted that it has enough resources or core competencies to support this. Maybe it is better to focus on two formats, such as supermarkets and hypermarkets, and to improve their core competencies.

Another problem is that Lianhua cannot achieve high management efficiency. In the last decade, its store number increased quickly while the profit margin did not increase accordingly. Lianhua plans to develop 6,000 stores by 2005, while its profit margin still targets keeping 1 per cent, quite low compared with the 3.6 per cent of Wal-Mart's Chinese stores. Without high efficiency, a retailer with a large number of stores is just like a dinosaur and will find it hard to succeed in the intensive competition when the protection policies are lifted in 2005. Since China entered the WTO, Chinese retailing has become more open and MNRs' expansion has accelerated. The time before 2005 is the most important for Lianhua, because the industry protections to Chinese retailers will be lifted after 2005. Lianhua should take full advantage of the protection period to grow strong enough to compete with foreign retail giants later. Therefore, taking an appropriate strategy is important. It can be argued that Lianhua needs the focus strategy: regional focus and retail format focus. Regional focus means Lianhua should focus on urban areas. Lianhua is still not able to compete nationwide. Focusing on urban areas can mean that limited resources are used centrally, which can make Lianhua easily achieve absolute advantages there. Retail format focus means that Lianhua should focus on one or two retail formats in the region according to local conditions. Its expansion model in urban areas may be organic growth and acquisitions while the JV style is not recommended. Meanwhile, the rural market may not be ignored. In rural markets, Lianhua may focus on Eastern and Southern China where they are richer than other rural markets by taking franchising and JVs for development; and the format there may be supermarkets. The aim of the focus strategy is to develop absolute advantages in market share and CA in urban cities, and to build strong brand influence with considerable market share in rural areas. Therefore, the key problem that Lianhua is facing is what kind of strategy it will take.

10.4.3 Lianhua's future

Lianhua's future is uncertain for two reasons. The first is that its 70 per cent of sales and 80 per cent of profit come from the Shanghai market, which has matured and has little potential for development. Meanwhile, many foreign retail giants such as Carrefour have been here. After China lifts restrictions on MNRs, it is doubtful that Lianhua can maintain its advantage in Shanghai. The other reason is that its joint ventures with Carrefour, six Carrefour hypermarkets in Shanghai, which Lianhua takes 45 per cent shares while Carrefour takes 55 per cent, are the most profitable. The six hypermarkets contributed 40 per cent Lianhua's pre-tax profit in 2002 and 30 per cent in 2003. It is very likely that in the near future, Carrefour will open its own hypermarket or kick off Lianhua when the restrictions on ownerships are lifed after December 2004. Lianhua's own hypermarket is over 30 per cent lower than Carrefour in sales per square meters. As such, the outlook for Lianhua competing directly with foreign retail giants such as Carrefour and Wal-Mart is not promising.

Another important factor infiuencing Lianhua's future is the formation of Brilliance Group. In April 2003, Lianhua became a part of Shanghai Brilliance Group, a behemoth created through a merger of stateowned Shanghai Yibai Group, Shanghai Friendship Group, Hualian Group and Shanghai Materials Group. With around 4,500 outlets under its control, Brilliance has a registered capital of RMB 1 billion (US$120 million), assets of RMB 33.5 billion (US$4.1 billion) and annual sales of RMB 80 billon (US$9.4 billon). The company plans to merge its 828 supermarkets, 11 hypermarkets and over 1,000 convenience stores under the Lianhua brand, expanding to 8,000 outlets nationwide by 2008. The strategy of Brilliance will directly influence Lianhua's development and future. Lianhua is facing a strategic problem: what strategy it should take in future development and in competing with foreign giants when China completely opens its retail industry to MNRs.

Although China presents tremendous opportunities for Chinese and MNRs, the fast changes in the Chinese transitional economy and Chinese retail industry ask both of them to keep learning and to keep examining their strategies so that they can change their operations according to Chinese market conditions to be competitive. Chinese retailers are in disadvantages in both size and core competencies; and have advantages in local advantages and understanding Chinese consumers. The future of Chinese retailers is uncertain, but it is full of promise if they can take right strategies.

10.5 Case study: Shenzhen Vanguard Supermarket Department Corporation Limited

Shenzhen Vanguard was established on 20th December 1991 as the "Shenzhen Vanguard Chain Commerce Ltd" invested by Shenzhen Wanke Enterprise Corporation Limited. Its retail format was the traditional department store. Because it could not develop its own CA effectively, its performance was disappointing: The two new stores closed in 1993, the same year opened, and the company had lost RMB 380 million Yuan by early 1994. On 17th January 1994, the company was restructured and renamed "Shenzhen Vanguard Super Department Corporation Limited Corporation." The holding company is still Shenzhen Wanke Corporation Ltd. Then Vanguard organized a team that went to American to study American retailing, and Wal-Mart's Sam's Club impressed them deeply. When they came back, they opened a similar warehouse on 17th July 1994. The low price image made the store achieve great success. The store was just 4,400 square meters, but its sales on the first day opened reached RMB 200,000 Yuan; the sales for the first half-year reached RMB 80 million Yuan. In the next three years, its annual sales were over RMB 330 million Yuan.

The background of retailing in Shenzhen: Shenzhen was the first Economic Special Zone in China as an economic reform pilot. It is an emerging immigrant city with about a 7 million population. The average age of the population is about 33 years old. The city has the highest disposable income per capita and the lowest Engel Co-efficiency in China. By 2000,

Shenzhen had 49 supermarkets with over 10,000 square meters for each one. The people mainly have the following characteristics in their consumption: they depend on supermarkets more than other cities because of small size families and limited shopping time; their spending on food tends to decline while in merchandise tends to increase; they like fashion and new lifestyle. Therefore, modern retail formats, such as supermarket, hypermarket and convenience store, are more popular in Shenzhen than in other Chinese cities. Both Wal-Mart and Carrefour opened stores there.

10.6 Vanguard's strategy and its general merchandise store (GMS)

Vanguard did not stay at its first success of the imitation of Sam's Club. After completely analyzing the market and its success experience, Vanguard decided to invent the General Merchandise Store, a new retail format combining the traditional supermarket and department store. This was a strategic decision. On 2nd November 1996, two months after Wal-Mart entered Shenzhen, Vanguard opened its first GMS, Cuizhu branch, to directly compete with Wal-Mart. The store was 12,500 square meters in size with over 60,000 categories of goods. Meanwhile, the store created an environment of entertainment and leisure. The new retail format met consumers' one-stop shopping not only in its wide selection of goods but also in its comfortable shopping environment. To the east of the store, just one kilometer away, was a Wal-Mart store of 18,000 square meters, and to the west, also one kilometer away, was another Wal-Mart store of 12,000 square meters. However, the sales of this Vanguard store were more than the sum of the two Wal-Mart stores. In 2000, the total sales of the store were over

RMB 460 million Yuan. Vanguard won the competition with Wal-Mart by its innovation. In the competition with Wal-Mart, Vanguard has kept obvious advantages and never lost the competition.

On 19th December 1998, Vanguard opened another 15,000 square meter GMS store, Caitian branch. Its goods reached 70,000 categories. The innovation in this store was its wide application of IT; there were many computers providing customers with store information and other public information and 96 large TV screens introducing products, doing advertising and providing public information. In 2000, the store's sales were over RMB 550 million Yuan. On 1st January 2000, another Vanguard's GMS was opened in the Baoan district of Shenzhen. The store is 27,000 square meters in size with about 100,000 categories of products.

On November 25th 2000, the fifth GMS of 22,000 square meters was opened with about 80,000 categories goods. This store explores branding general merchandise, especially in clothing, which targets the M Class by its unique brand image and is operated like an apparel specialty store within the GMS store. Since the store is large, Vanguard pays attention to creating a comfortable shopping environment, which is concerned greatly with local consumers. With its 5 stores, Vanguard achieved sales of over RMB 1.6 billion Yuan in 2000, ranking first place in Guangdong Province and the 13th largest retailer in the nation. Its average sales per store were about RMB 320 million Yuan, higher than Lianhua's, the largest retailer in China, whose sales per store were about RMB 106 million Yuan, and Carrefour's RMB 296 million Yuan per store.

Vanguard's success has firstly resulted from its unique retail format, GMS. GMS has the advantages of both the department store and the supermarket, and meets consumer demand for "one-stop shopping" better than the supermarket and hypermarket by its wider and deeper assortments. Supermarkets' profits mainly come from the food section. The fierce competition has squashed the profit margin of the food section greatly while merchandise sold in department stores normally has a high profit margin. Further, in Shenzhen the Engel Co-efficiency is the lowest in China and people there spend more on non-food products than in other cities. GMS can attract customers from both the supermarket segment and the department store segment, so its customer segment is larger than either of the two. Among its sales, 45 per cent are from the department store section. In this way, Vanguard not only avoids direct competition with supermarkets but also uses the department store sector as an extra advantage. The wider and deeper product assortments and broader price range make Vanguard meet customer demands better than Wal-Mart and Carrefour. Wal-Mart and Carrefour take the strategy of 80/20, which means 80 per cent of their profit comes from 20 per cent of the goods. This needs accurate information in selecting the 20 per cent goods. However, under the conditions that it is very difficulty to obtain complete and accurate information in China and the market changes quickly, a narrow and too selective assortment has high market risk. Vanguard overcomes this difficulty by providing more choices than Wal-Mart and Carrefour. Wal-Mart's price range is from RMB 0.10 Yuan to RMB 30,000 Yuan while Vanguard's range is from RMB 0.01 Yuan to RMB 90,000 Yuan. The broader price range provides customers with more choices, which attracts more customers for shopping. In procurement, Vanguard takes its unique centralised procurement. It allies with its suppliers and shares profits with them, which stimulates the suppliers to provide Vanguard with the lowest price. In its expansion, Vanguard takes the saturation strategy. It believes that by opening 5 stores in one city it is much easier to succeed than by opening 5 stores in 5 cities separately. Further, the strategy also makes Vanguard benefit fully from local advantages and easily achieve economies of scale. The successful experiences of Vanguard are called the "Vanguard Model," which can be summarized by its GMS format (supermarket plus department store in operation) and saturation growth strategy focusing on the local market.

After succeeding in Shenzhen, Vanguard decided to enter its neighboring cities. On 1st December 2001, it opened the first GMS in Guangzhou. The 13,500 square meters four-floor store has over 60,000 categories of goods, about 10,000 of which are specially selected according to local residents' tastes. The store provides "one-stop shopping" by combining shopping, entertainment and leisure. It is trying to promote its GMS in Southern China.

10.6.1 Acquisition and the great changes in corporate strategy

In August 2001, China Resources Holdings Limited (CRH) acquired a 65 per cent stake of Vanguard by RMB 457 million Yuan; on 29th October 2001 it changed Vanguard's name to China Vanguard Super Department Co., Ltd. CRH is a Hong Kong based Chinese SOE. It was founded in 1948 and its total assets reached HK$56 billion in 2001 with over 90,000 employees. Its China Resource Enterprise Limited (CRE), which was established in 1992, is a floated company on the Hong Kong stock market and one of the largest Chinese SOEs in Hong Kong; its business scope covers many industries including oil, machinery, pharmaceutical, food, textile, etc., with over 66,000 employees. In early 2002, CRH acquired the remaining 28 per cent of the shares and Vanguard was completely acquired by CRH. On 1st July 2002, CRH merged

Vanguard and CRH's supermarket sector; the integrated entity is named by China Resources Vanguard Supermarket Corporation Limited (CRV). The main reason for changing the names is for future national expansion. On 1st August 2002, CRV was under CRE. By the end of 2002, CRH held 55.47 per cent of CRE's shares and 35 per cent of CRV's shares; CRE held 65 per cent of CRV's shares. Besides, CRH also acquired 39.25 per cent shares in the Suguo Supermarket by RMB 232 million Yuan in September 2002. CRV's store numbers reached 400 by 2002. However, after the acquisition, CRV faces great changes:

1. *Changing the corporate strategy*: CRH and CRE asked CRV to change the former Vanguard Model. They claim to make the "Chinese Wal-Mart" and propose the "Four-Five Project" for CRV, which is to invest RMB 5 billion Yuan within the next 5 years to realize sales of RMB 50 billion Yuan by 2006 with an annual profit of RMB 500 million Yuan. To realize the project, CRV has to change from a regional player to a national operator and to make a new strategy called: "operating across regions with multiple-format and expansion by over-normal speed with sustainable optimizing operation," which is targeting the three markets: Eastern China, Southern China and Northern China, developing four retail formats in the markets: GMS, discount store, supermarket and the standard supermarket, and expanding fast.

2. *The Change of management team*: The acquisition brings great changes in CRV's management team. On 14th January 2003, the General Manager of CRV, Mr. Xu Gang, who was the CEO of the former Vanguard and the main inventor of the Vanguard Model, resigned from his position, which shocked the industry and was called "the biggest news in Chinese retailing." (China Business, 7th March, 2003). Mr Xu is often called by "the first person in the Chinese retail chain" and "the soul of CRV."

The direct reason for Mr. Xu's resignation was the discrimination of CRV's salary policy. The management team of CRV came from three sources: some were from the former Vanguard; some were from the former retail sector of CRE; and some were from other companies of CRE who do not understand retail business. The managers from the latter two sources did not like to be managed by Mr. Xu. They often did not co-operate well with his work and showed arrogant attitudes in their work. Further, CRE paid the managers at CRV on different salary systems, although they all worked in CRV; the employees from the former Vanguard obtained lower salary levels than those from CRE and sometimes they even could not be paid in time. However, the deeper reason for Mr. Xu's resignation was due to the integration problems between acquired Vanguard and the CRE. After Vanguard was acquired, its management team led by Xu Gang, which was called the golden team managing Vanguard for years, was broken down. Although Mr. Xu Gang was appointed as the General Manager of CRV, he lacked effective support in his work. He had four vice General Managers, only one of them was from the former Vanguard; among the other three, only one knew retail business well and was asked to manage the finance department. In the senior management team of CRV, only 5 managers were from Vanguard. Further, Mr. Xu Gang did not agree to CRV's fast expansion strategy, particularly the "Four-Five Project"; he still preferred his Vanguard Model and his saturation and gradual expansion strategy. Besides, he felt it hard to adapt to CRV/CRE's corporate culture, which was very bureaucratic and very different from Vanguard's. After Mr. Xu resigned, another 19 managers from the former Vanguard also resigned, accounting for one third of CRV's senior management team.

CRV's performance and its future. Led by Mr. Xu Gang, CRV ranked the largest retailer in Southern China and the 7th largest in the nation with annual sales of RMB 8.59 billion Yuan in 2002. However, the fast expansion strategy soon met problems. Since Vanguard was acquired in 2001, the Vanguard GMS was copied by CRV at a surprising speed. For example, in Southern China, from September 2001, CRV had opened 4 GMS in three cities: Guangzhou, Zhuhai and Zhongshan. During Vanguard times, led by Mr. Xu Gang, Vanguard opened only 6 GMS within six years from 1994 to 2000 in order to guarantee that each store opened succeeded. The performances of the newly opened stores cannot meet the expectation actually; they are facing many problems:

(1) *The shortage of human resources*: The lack of qualified management personnel has become the bottleneck for its fast expansion. To open one new store, it needs at least 4 or 5 store managers, 23 category managers, and 80 team leaders. The demands for these management personnel increased four times within one year in 2001 reaching 1,283. The current management teams have been not enough to support the rapid expansion. The company is trying to solve the problem in several ways:

- *Promotion*: The best way to solve the problem is to promote managers from the current employees, but the available human resources are limited.

- *Employing from outside*: Because of the intensive competition, excellent managers are scarce and targeted by each retailer. Employing from outside is expensive.

- *Training*: An appropriate way to solve the problem is by training. But it takes a long time to train qualified managers from employees, particularly from new employees.

- *Transfer*: CRV can transfer some managers and employees from current stores to new stores. However, for a large number of front-line employees, it has to employ from local markets. Some new employees could not adapt to the corporate culture or hard work and left the company. In the store of Zhongshan city, within the first three months, over 390 employees left for the corporate culture reason.

(2) *Problems in the operation*: For example assortment management, display management and merchandising management, etc. The worst is the problem of availability. In some of its newly opened stores, the shortage of commodities has even reached 20 per cent; in other words, 20 per cent of the commodities in its stores are often out of stock and cannot be supplied in time. This reflects problems in its relationships with its suppliers.

Facing these problems, CRV has to slow down its expansion. Managers find that the successful model developed in Shenzhen does not work well when it is transferred to the new markets, such as Guangzhou, Zhongshan and Zhuhai, because of different consumption cultures and consumer behaviors. Localization is difficult even for Chinese retailers. A manager complains that "Zhongshan is just 2 hours journey from Shenzhen, but it is like another world and has so much difference in consumer behaviors." The same problems also exist in other entered cities, although they are near Shenzhen.

Therefore, CRV has lost money since January 2003, although it was profitable in 2002. It lost RMB 15 million Yuan in January 2003 and about RMB 20 million Yuan in February 2003. Actually, all its GMSs of former Vanguard are still profitable while all supermarkets of the former CRE retail sector are losing money. Before they were merged, the annual sales of the six Vanguard GMSs in Shenzhen were over RMB 1 billion Yuan with more than RMB 60 million Yuan net profits. In 2002, over 400 supermarkets of the former CRE lost about RMB 20 million Yuan, but because the GMSs were profitable, the whole performance of CRV was profitable. However, GMS also meets problems: all GMS stores opened after October 2002 have lost money.

After Mr. Xu Gang left, Mr. Chen Lang, who has no retail background, was appointed as the General Manager of CRV. Facing this situation, CRV has to slow down in its expansion and make some adjustments. It will gradually give up the GMS format and focus on the other three formats. The former Vanguard GMSs will be changed to "supermarket + fresh foods and vegetables + furniture + clothing" becoming the neighborhood shopping center. CRV has enough capital for its future expansion supported by the floated company. The key problems that CRV is facing are that it needs an appropriate strategy and a powerful management team. It can be argued that acquiring the management team of Vanguard is more important than acquiring the stores. If the former Vanguard's management team is allowed to lead the CRV, the result may be different. But seen from the current situation, the future of CRV looks worrying.

10.7 The case of Suguo Supermarket Co., Ltd: developing the Chinese rural markets

Since Vanguard succeeded in Shenzhen, some retailers are marching to Chinese rural markets, a vast and great potential market. Competition in urban cities is becoming fierce, while the rural market seems quiet. Although it has nearly an 800 million population, many retailers still ignore the market, because the urban markets are still far from saturation. However, a local retailer, Suguo Supermarket Ltd (Suguo), is writing a success story there by its excellent performance. Suguo was established in July 1997 in Jiangsu Province by the Jiangsu Provincial Supply and Marketing Cooperative. By 1999, it had developed from 1 store to 130 stores, but mainly concentrated in Nanjing City, the capital of Jiangsu Province. Compared with other local retailers, it did not have obvious CAs over them. Then it put its eyes on rural markets, where it had unique advantages because of the history. The Supply and Marketing Cooperative was responsible for distributing commodities in the rural areas and was the dominant retailer there with the largest networks. After market research, Suguo found the rural markets had the following characteristics:

- Huge consumer group. Rural market had about 800 million people.

- Low revenues compared with urban residents although their consumption value was not so much different from urban consumers, especially to those who were in suburban areas.

- Comparing with urban markets, rural customer demands were more easily satisfied than urban consumers, with less developed retailing.

- Except for the cooperative stores, most rural retailers were private-owned independent small stores.

- Rural consumers actually paid more than urban consumers for the same commodities; they often face smaller selection due to fewer goods being supplied.

It found its advantages and capacities were:

- Jiangsu Supply and Marketing Cooperative had 48,000 outlets in the rural market.

- Nearly all employees of these outlets were from rural areas, which made them know rural customer demands well.

- These outlets were all independent operators and had lost money for years because of the poor selection of commodities with high price.

- These independent retailers did not have much capital for further development.

- Suguo had a good brand image and reputation in Nanjing city and had been well known by the rural customers within the province.

Then Suguo decided to enter the rural market and to restructure the cooperative networks by a franchising operation. On 26th April 1999, its first franchise store in Lishui County was opened. In the past, this store just had about 2,000 categories of products and its sales were just between RMB1,000 Yuan to RMB 2,000 Yuan each day. It increased its commodities to 12,000 categories. On the first day of opening, the sales reached RMB 70,000 Yuan. This success presented a good model to other local retailers. They asked to join the franchise with great enthusiasm. Within just two years, Suguo's franchise stores had reached 450.

Sun Feng, the manager of a small store at the town called Tang Shan near Nanjing city, told his story. Two and half years before, his store had lost RMB 1.7 million Yuan and had been nearly out of business because of the limited selection of goods with poor qualities and high prices. Then he chose to join Suguo and became one of its franchisees. He received training from Suguo and all goods were supplied by Suguo. With rich selections, good quality and cheap prices of commodities, his business was booming with over RMB 5 million Yuan sales each year. The sales per day increased from several thousand Yuan to tens of thousands of Yuan, and on 9th January 2002, its daily sales reached RMB 180,000 Yuan.

During its franchising operation, Suguo has developed a complete management system and franchise conditions, which mainly include two parts: requirements in property rights and conditions in management. In property rights, Suguo asks the franchisees (the former cooperative store operators) to reform their property right structures, changing from the former collective-owned to either private-owned by management buyout or to limited equity companies. By this reform, the former employees became the shareholders or owners of the stores, which greatly improves their motivation and efficiency. In management, it takes "Seven Central Managements," which are central procurement management, central distribution management, central pricing management, central financial management, central standardized operation management, central developing management and central image management. In order to take merchandising management, it applies a POS system in each rural store. Suguo also builds an efficient supply chain system. Its large distribution center is 65,000 square meters and is able to distribute 25,000 category commodities directly to its stores. Through this, it distributes about 40 per cent of its commodities. Those commodities that it cannot distribute are supplied to the stores from fixed suppliers in order to guarantee the quality of goods. In addition, Suguo also develops PL products by cooperating with suppliers for specially servicing rural consumers. By 2001, it had developed 670 stores including 450 franchises and 220 directly invested; the total sales reached RMB 5.28 billion Yuan, ranking 7th among the national chain stores. It targets rural consumers and over 70 per cent of its stores are in towns and counties and over 50 per cent of its sales are from rural markets. It is entering its neighboring provinces, such as Shangdong, Anhui and Henan. Suguo was the first Chinese retailer to develop chain operation in the rural market.

Suguo is not the only case in successfully developing Chinese rural markets. In April 2002, the Supply and Marketing Cooperatives of Zhejiang Province, which had 5,473 store outlets, cooperated with Shanghai Hualian, the second largest Chinese supermarket chain, to develop the rural markets in Zhejiang Province, where the urban market is saturated while the rural market is still less developed. The purchasing power in some rural markets is similar to that of the urban market. In future, sooner or later, the rural market will be a vital battlefield for determining who will be the winner in competing Chinese retailing.

10.8 Summary

Most Chinese retailers' business model is channel fee based, which is just a simple retail trader model. They make profits by charging channel fees from their suppliers. The Chinese channel fee is not just one kind of fee but a group of fees including many different fees and varies from one retailer to another depending on a supplier's bargaining power; so does the amount of the fee. The over-charged channel fee brings serious negative impacts on developing the Chinese retailing, particularly weakens the competitiveness of both Chinese retailers and Chinese retailing. Some Chinese retailers are exploring competitive models. Most Chinese retailers, still have a long way to go.

Competition between foreign and Chinese retailers in China

11.1 Competing with MNRs

What does competition mean to retailers? What do retailers compete for? It can be argued that the answers are the same: customers. Retailers compete for customers. Those retailers who win customers will succeed in the competition while those who lose customers will fail. In order to win customers from its competitors, a retailer has to develop its CA over its competitors. Both Chinese retailers and MNRs have their own advantages and disadvantages in developing their CAs.

11.1.1 Chinese retailers' advantages and disadvantages

The main advantage that Chinese retailers have can be called the local advantage, which could be defined as the capability to take full advantage of local resources and to deeply understand local markets, particularly consumer demands in order to develop their local markets effectively. Many Chinese retailers have great strength in this kind of local advantage. They have rich knowledge on their local markets and control a wide range of good store locations; they are very familiar with local consumer behaviors and local tastes; they have good public relations with their suppliers, local governments and stable distribution networks for their long-term operation; they have strong local brands with considerable loyal customer groups; they are good at dealing with all kinds of unexpected incidents influencing their normal operation, etc. Supported by the local advantage, Chinese retailers are able to compete with MNRs.

This local advantage is further strengthened by the conditions of Chinese retailing, which prevent MNRs from transferring their successful experiences obtained abroad to China. For example, MNRs are good at and have got used to central management and central sourcing. While in China, chain operation is difficult for both policy reasons and operation difficulties, such as the fact that the Chinese logistics industry is undeveloped and very fragmented and thus the transportation cost is expensive; products have to be mainly sourced in local markets, while their quality is hard to control because different materials are used by different suppliers. Thus some advantages of MNRs are weakened. Actually, it may be better for MNRs to treat China as several markets rather than just one market. The General Manager of Guangzhou Dingyi Food Company, a Taiwan-based company, Mr. Lai said, "The first rule of developing business in Mainland China is to treat it as over 30 different countries." Chinese retailers with local advantages, also face the problem of localization during their expansion because of diversified local cultures and local tastes. On the other hand, Chinese retailers also have many disadvantages compared with MNRs: most of them are on a small scale, have weak management and lack capital as well as competitive models. No Chinese retailers had built a complete Enterprises Resource Planning (ERP) system and CRM system by 2002. Most of the Chinese retailers have very good decoration in their stores, but their core competence is weak. Many local retailers, particularly

SMRs, are going under in the competition. But after several years' competition with foreign retail giants, some local retailers are becoming stronger and more experienced. Carrefour now does not have more CAs in competition, particularly in price, in some cities, such as Chongqing, Wuhan, Shenyang and Shenzhen. Many local retailers are growing up and developing their competency quickly by their excellent learning ability.

11.1.2 MNRs' advantages and disadvantages

MNRs have many advantages in their home countries. But some of them cannot be transferred to China, particularly their advanced logistic systems and distribution systems. The main Wal-Mart's CA is its EDLP, which is based on its complete and advanced supply chain system built in the States; while after it entered China, the system cannot be transferred because of the less developed infrastructure conditions and the restriction of Chinese policy. In China, the first advantage that MNRs often have is their brand names. A well known and famous reputation is very important in a retailer's internationalization. The next advantage is their strong financial strength, which is a main weakness of Chinese retailers. MNRs are normally able to afford losses for a long time in a price war. In addition, their advantages also include excellent management skills, wide application of IT, such as the MIS and ERP systems, strong bargaining power in procurement, unique skills in inventory control, etc. My research finds that in China, foreign companies do not outperform Chinese retailers in gross margin, but they have an absolute advantage in their inventory management, which greatly contributes to their profit (Table 11.1).

Table 11.1	The performance comparison between Chinese retailers and MNRs	
	Chinese retailer	MNRs (in China)
GMROI	0.72	1.67
Gross margin	12 per cent	11 per cent
Inventory turnover	6	15.2
Store inventory turnover	45 days	24 days

Note: GMROI is gross margin return on inventory investment; it measures how many gross margin dollars are earned on every dollar of inventory investment.

The main disadvantage that MNRs have is their lesser understanding of the Chinese market and their weaker capability in localization than Chinese retailers. Whether a foreign retailer can succeed in China mainly depends on if it can successfully realize localization. When Wal-Mart first entered China, product assortments, particularly food assortment, made its performance disappointing, although it had taken two years' market research before it opened the first stores. The case of Park' N Shop also gives convincing evidence of the difficulty of localization. Another disadvantage that MNRs have is that they are too large, which often makes them inflexible and very bureaucratic. During the competition between a local retailer, Renrenle, and Wal-Mart in Shenzhen, the more flexible management in Renrenle made the company able to react to market change in time, particularly during the price war with Wal-Mart; while Wal-Mart's strict and large management system reacted slower than Renrenle, which made it hard for Wal-Mart to win over Renrenle. MNRs may have another disadvantage: They have been too successful before. This makes them get used to following the successful experience gained in their home markets, which may not work in China. The models and strategies they have developed particularly for their home markets often do not suit the Chinese market because of the different consumer demands and different Chinese conditions for retail operation. To succeed in China, MNRs need innovation rather than just copying their former successful experiences.

After entering China, many MNRs just regard other MNRs as their main competitors while they ignore the threats from local retailers. The lesson has been serious. The old Chinese says "a powerful dragon cannot crush a snake in its old haunts; even a powerful man cannot crush a local bully." That's why in Chongqing City, local retailers took 90 per cent of the market share while Carrefour just ranked at the fourth position there by 2001. Carrefour and Wal-Mart have learned many lessons from the competition with Vanguard and Renrenle respectively in Shenzhen. In the competition, it is possible for Chinese retailers to win over foreign retail giants if Chinese retailers take the right strategies and use their advantages well.

11.2 The evaluation of Chinese retailing before accession to the WTO

Although it grew quickly before China's accession to the WTO, Chinese retailing was inefficient and not competitive as a result of the past Chinese economic system and the history of Chinese retailing. In the past decades, Chinese retailing developed independently in a closed environment with little influence from the global retail market. The whole industry was very fragmented and presented a clear dual-structure. The developed urban retail market and the less developed rural market coexisted; and modern retail formats and traditional markets also coexisted. The majority of Chinese retail sales were from a minority part of the 495 million urban population rather than the national 1.3 billion population. Although all modern retail formats had emerged in China within the 10 years from 1992, they were only at the beginning stage and had not yet developed successful operation strategies. Supported by the Chinese government, chain operation had become a trend, but the scale was still small. Although the emergence of private-owned retailers broke the monopoly of state-owned retailers and brought competition in Chinese retailing, they were weak and it was still SORs that dominated the industry, while SORs were facing many problems, such as low efficiency, triangle debts and weak competency. Chinese retailers were less competitive, and few of them had developed both IF and OF. Their supply chain system was inefficient compared with MNRs, their brand images were only local or regional at most, and they were very weak compared with foreign retail giants.

In 2001, the total sales of the top 100 Chinese retailers were RMB 234.2 billion Yuan (about US$28.2 billion) accounting for 5.06 per cent of the national sales, while Wal-Mart's sales were US$220 billion accounting for about 9 per cent of American retail sales. This indicated that Wal-Mart's sales were about 28 times those of the 100 largest Chinese retailers' total sales. Wal-Mart's annual sales in 2001 were 130 times of Lianhua's. Chinese retailing was less competitive compared with retailing in developed countries because of the unfavorable input factors and less developed supporting industries, such as the factor conditions, demand conditions, related and supporting industries, and intense local rivalry.

Factor conditions. Most Chinese retailers were facing a serious shortage of qualified management personnel, such as professional purchasing managers, store managers, inventory managers, etc. and the lack of enough capital for development. Obtaining loans from banks was costly and difficult because their debt ratios were high. Most Chinese retailers operated with over 70 per cent debt ratios. Capital had become the bottleneck preventing retailers from further expansion. In addition, there were few university research programs and university degree programs in retailing. Few national research institutes played an important role in the retail sector. Most Chinese retailers were SMRs and lacked the application of IT. Few Chinese retailers had built MIS, CRM and ERP.

Related and supporting industries. Few Chinese retailers had developed partnerships with their suppliers; the unstable relationships with their suppliers such as the conflicts caused by charging "channel fees" greatly weakened their competitiveness. The Chinese logistic industry was very fragmented, inefficient and undeveloped; the distribution of goods over 500 kilometers was often difficult; the supporting industry was very weak.

Chinese home demand. The Chinese consumption level was still low compared with developed countries. They were unsophisticated customers. Their high Engel Coefficient indicated that Chinese consumptions mainly focused on basic goods. But their demands were growing fast benefiting the sustainable growth of GDP.

Local rivalry. Rivalry was increasingly intensive but fierce competition was just on price. Most Chinese retailers copied foreign players; and the government still protected some industry from foreign competition. Many international retail giants had entered the industry and obtained "over national treatments" provided by some local governments; they had developed their CAs by adapting to local markets.

Therefore, before China entered the WTO, the critical input factors in retailing were of poor quality or unfavorable; critical supporting industries were very weak and customer demands were still not sophisticated, although some of the competition was excessive, it was pervasive imitation and there was an absence of distinctive positioning. The main favorable factor was that the future prospect of the market was promising. Since Chinese retailing was in transition, many opportunities were emerging from the following issues:

- Sustainable economic growth, which produced an enlarging M Class.

- Revolution and evolution of different retail formats. Traditional formats were evolving into modern formats. All modern formats were at the very beginning stage. The potential for their further development was huge.

- Shifting in consumer demands and changing consumer behaviors.

- The capacity of the Chinese market. Most MNRs just operate in big cities in Eastern and Southern China; there would be a vast market in other regions for exploration in future.

- The emergence and growing of new consumer segments, such as the S generation provided retailers with huge opportunities.

Challenges always follow opportunities. These challenges mainly came from:

- Different consumer patterns. China is large. Each region has its own consumer pattern. A successful model developed in one place may not work in another.

- Competition from local retailers. Although there had been no real national champion in China, there were huge numbers of local or regional champions with strong local advantages.

- Undeveloped logistics industry and infrastructure, which made operation costs high, particularly to MNRs. They would have to take much time to adapt to the conditions and tailor their successful models developed from other markets.

- Local protectionism, which prevented chain operation and fair competition.

11.3 Summary

As an industry, Chinese retailing was uncompetitive; but that does not mean Chinese retailers are not able to compete with MNRs. The case studies suggest that Chinese retailers are not only able to compete with MNRs but also out perform them in different ways. The Lianhua case suggests that Chinese retailers can increase their core competencies and grow fast by using the right strategies, particularly the alliance strategy taking advantage of various kinds of resources. The Vanguard case suggests that Chinese retailers can win the competition with foreign retail giants by their innovations. The Suguo case suggests that Chinese retailers can also succeed by avoiding direct competition with foreign retail giants and concentrating on the vast rural market. Facing the WTO, both Chinese and MNRs would meet challenges and opportunities that they never meet before. Chinese retailing had great potential to catch-up on the eve of the WTO. If Chinese retailing can take advantage of the WTO and its own local advantages, it is completely possible for them to catch up. Chinese retailing was standing at the critical moment in its history on the eve of the WTO.

Part IV
The WTO and Chinese Retailing

WTO: a great opportunity to fail or succeed?

12.1 The impact of China's accession to the WTO on Chinese retailing

In December 2001, China formally became a member of the WTO. In the accession agreement, the Chinese government made a number of promises with respect to retailing,

Foreign service suppliers may supply services only in forms of joint ventures in five Special Economic Zones (Shenzhen, Zhuhai, Shantou, Xiamen and Hainan) and six cities (Beijing, Shanghai, Tianjin, Guangzhou, Dalian and Qingdao). In Beijing and Shanghai, a total of no more than four joint venture retailing enterprises are permitted respectively. In each of the other cities, no more than two joint venture retailing enterprises will be permitted. Two joint venture retailing enterprises among the four to be established in Beijing may set up their branches in the same city (i.e. Beijing).

Upon China's accession to the WTO, Zhengzhou and Wuhan will be immediately open to joint venture retailing enterprises. Within two years after China's accession to the WTO, foreign majority control will be permitted in joint venture retailing enterprises and all provincial capitals, Chongqing and Ningbo will be open to joint venture retailing enterprises.

Foreign service suppliers will be permitted to engage in the retailing of all products, except for the retailing of books, newspapers and magazines within one year after accession, the retailing of pharmaceutical products, pesticides, mulching films and processed oil. Within three years after accession and retailing of chemical fertilizers within five years after accession.

None, within three years after accession, except for: retailing of chemical fertilizers, within five years after accession; and those chain stores which sell products of different types and brands from multiple suppliers with more than 30 outlets. For such chains stores with more than 30 outlets, foreign majority ownership will not be permitted if those chain stores distribute any of the following products: motor vehicles (for a period of five years after accession at which time the equity limitation will have been eliminated), and products listed above and in Annex 2a of the Protocol of China's WTO Accession.

The foreign chain store operators will have the freedom of choice of any partner, legally established in China according to China's laws and regulations (Source: Report of Working Party on the Accession of China, WT/ACC/CHN/49/Add.2 1 October 2001).

The accession agreement indicates that Chinese retailing will have three years' transitional period from December 2001 to December 2004, during which MNRs still face the limitations in their chain operation, ownership, operating region and business scope; after December 2004, nearly all these limitations will be lifted.

It can be argued that the accession will impact on Chinese retailing by two ways: the direct impact and the indirect impact. The direct impact comes from the further opening of Chinese retailing itself caused by the accession, such as more foreign entrants and the faster expansion of the existing foreign entrants in the market after the accession (Figure 12.1). The indirect impact is a result of the changes of other Chinese industries caused by the accession, such as the restructuring of the Chinese logistics industry, the lower import tariffs, the changing lifestyle of Chinese consumers, etc. It can be argued that in the short term, such as in the five years after the accession, the direct impact will be strong while the indirect impact will be weak. However, in the long term, the direct impact will be weak while the indirect impact will be strong.

In the short term, the accession will make the retail competition in the Chinese market more intensive. For example, it can bring an earthquake effect in some Chinese cities due to more foreign retail entrants and the accelerating expansion of the existing retailers in the market. My research finds that in Shenzhen, when Wal-Mart just opened its stores, the sales of Chinese stores decreased by about 10 per cent on average. However, in the long run, this direct impact tends to gradually become weaker because of the fast growth of Chinese retailers and the improvement of their competencies during their competition with MNRs. In Shenzhen, some local retailers, such as Vanguard and

Figure 12.1 The impacts of China's accession to the WTO on Chinese retailing

	Short-term Impact	Long-term Impact
Direct Impact	strong	weak
Indirect Impact	Weak	strong

Renrenle, grew up quickly during the competition with Wal-Mart and Carrefour; Wal-Mart and Carrefour have already had no significant advantages over them in the competition there. Further, the competition among retailers including both MNRs and Chinese retailers will gradually change from only concentrating on the store operation to concentrating on the supply chain oriented operation, competing with all the supply chains, which is closely related to the Chinese logistics industry and retailers' distribution operations. It can be argued that retail competition among retailers is to a great extent competing for their distribution systems.

Before China's accession, the Chinese logistics industry was much undeveloped, and it was not open to foreign companies. Thus, MNRs could not fully perform one of their most important CAs, the highly efficient supply chain, in the competition; while the undeveloped supply chain is just the weakest side of Chinese retailers. Therefore, Chinese retail competition mainly focuses on competing for the front-line operation rather than for the back-line operation and for the whole supply chain. The unopened logistics industry protects Chinese retailers from the threats from MNRs. The opening of the Chinese logistics industry will benefit MNRs. Some MNRs can bring their logistic partners in their Chinese operation, which will make these MNRs transfer some of their FSAs to China by establishing their distribution systems in China. Further, the opening of the Chinese distribution field will bring a profound change in Chinese retailing. Distribution is critical for retail success. It is only when MNRs build up their complete distribution systems in China, can they develop complete supply chains, transfer and develop their CAs fully there. The accession will drive the restructuring of both the Chinese logistics industry and the distribution systems of retailers operating in China.

Thus the indirect impact from the Chinese logistics industry will be strong. In addition, it can be argued that any impact on the retail process in Chinese retailing and its related industries caused by the accession will also impact on Chinese retailing. For example, according to the GATT, China agrees to cancel all trade quotas and cut tariffs to around 15 per cent by 2005, which may result in a great increase of Chinese import commodities. This can bring negative impacts on Chinese retailers while benefiting MNRs: MNRs are able to take advantage of their global sourcing networks to import more foreign goods and sell them in their Chinese stores, which supports them in developing a differentiation strategy. Meanwhile, no quota control may make MNRs import more foreign products, which also provides them with stronger bargaining power over Chinese suppliers who supply them with similar products. On the other hand, the accession also makes MNRs increase their procurements in China and sell them in their international outlets, which also increases their bargaining power over Chinese suppliers, too. Further, according to the GATS, all MNRs will be offered national treatments. MNRs thus can compete equally with Chinese retailers equally. Their global operation systems such as global sourcing systems will benefit their competition in China. In another words, they can use their global resources to compete in the Chinese market. Therefore, the accession will bring MNRs many advantages and opportunities. Another important indirect impact is the Chinese regulations on mergers and acquisitions. On 4th November 2002, China Securities Regulatory Commission, the Ministry of Finance and the SETC jointly issued the regulation that allows foreign investors to use the public-tender process for buying Chinese state-owned or institutional shares in domestic-listed firms. Because many large Chinese retailers are listed companies, the impact could be great in future. With the Chinese government lifting its restrictions on M&A, it is becoming MNRs' main way to develop in China.

Combining the direct impact and indirect impact together, it can be argued that the impacts of the accession on Chinese retailing will be stronger in the long term than that in the short term. The structure changes caused by the accession could be analyzed by Michael Porter's Five Forces Model.

12.1.1 The five forces model analysis on Chinese retailing

- Chinese customers' bargaining power will become stronger, because the Chinese market has been a buyer's market and they will face more retailers and more choices in their shopping than ever before.

- The bargaining power of Chinese suppliers will be further weakened because: (1) the lifting of import quota control and the lower tariffs for imported goods will cause more foreign products to enter China; (2) MNRs can fully perform their advantage of global sourcing and use it to strengthen their bargaining power; and (3) their large quantity of procurement volume can reduce the suppliers' profit margin further.

- The threat of potential entrants is increasing, because both multinational and Chinese retailers can enter the industry more easily than ever before.

- The threat of substitutes is also increasing, because MNRs can enter Chinese retailing with both physical stores and cyber stores. The undeveloped Chinese e-tailing may be lost to foreign operators.

- The rivalry among the existing retailers will be more intense due to more entrants and their accelerating expansion. According to the accession document, China is lifting a series of restrictions on JV partnerships and geographical locations for MNRs, raising the prospect of stiffer competition.

Therefore, from Michael Porter's model, it can be argued that the barriers to entering Chinese retailing will be greatly lowered; the competition will be more intensive, which can reduce the profit margin of the whole industry to a narrower level, and the accession will drive Chinese retailing to evolve faster.

12.1.2 The impact of China's accession on different retail formats

Although the accession will bring Chinese retailing a deep influence, the influence varies across different retail formats.

Department store. The Chinese department store is facing two key problems: how to reposition itself and how to develop its differentiation advantage. Many of them are competing with supermarkets, because they target the same segment and take similar strategies to Chinese supermarkets. So price wars among the department stores happen frequently. In Shanghai, the most developed commercial city in China with about 700 department stores, one-third of them lost money in 2001, but price wars among the department stores still happened frequently. Chinese department stores need repositioning; they should not compete with the supermarkets and focus on price wars. It can be argued that they should target up-market. With rising incomes, more and more people join the M Class. They naturally tend to trade up from low-price types of retail outlets, and the department store has been the natural place to go, which has great prestige and glamour and has built much of its reputation around service in China. Therefore, the Chinese department store should be a downtown player and concentrate on image quality and service becoming an up-class operator. The accession gives the Chinese department store a heavy blow and its dominant status in the nation will be lost. The more intensive competition caused by the accession will further reduce its market share and accelerate its diversification. Some of them will be out of business, some may be acquired for their valuable locations, and some will be transformed to other retail formats such as specialty store, supermarket and GMS followed the old saying "If you can't lick, 'em, join 'em".

Supermarket. This is the largest number of retailers behind the department store in China. Most of them have evolved from the former grocery stores or department stores. Their main CA is their locations. They have controlled nearly all golden areas in Chinese urban cities. If MNRs want to enter this field, the best ways for them is to acquire local retailers or make JVs with them. Although foreign supermarket chains have had fewer successful cases than foreign hypermarket chains in China, some MNRs are now riding WTO liberalizations to boost their presence in China. For example, Aeon Co, one of Japan's top three supermarket chains, announced in late 2001 that it planned to open nearly 60 stores in the booming coastal regions of China by 2006. However, Chinese retailers must control the majority shares in JVs by December 2004 if the number of chain stores is over 30. Some MNRs have been out of the business in this field. Therefore, in this format, Chinese operators will dominate the market and the impact of the accession can be limited in the short term.

Convenience store. This format will be a main field that MNRs compete for. In China, many convenience stores are small-sized supermarkets and located in residential areas. Actually, they are not convenience stores by the modern concept. The number of this kind of retailer is limited while the demand is strong, which provides a great potential for future development. Most current Chinese convenience stores have evolved from former grain stores. Their main advantage is their free or low rents supported by local governments and other favorable policies from the governments. For example, to defend the competition from MNRs, the Shanghai government has decided to launch a convenience store project, which is to reform about 4,000 small neighborhood stores and change them to convenience stores by chain operation under the brand: "Easy Buy Super Convenience Store". The most difficulty thing for MNRs to enter the field is that they will take a long time to build their supply chains and distribution systems. Some Chinese retailers have entered the field but their CAs are quite weak. The success of CS largely depends on if the operator can successfully develop a highly efficient supply chain supported by IT, as 7-11 has done in Japan. Foreign operators such as 7-11 may dominate this format by their franchising operation and rich experience.

Hypermarket. MNRs have controlled this format in China, and they are very likely to continue dominating the format in future. Chinese retailers have local advantages while MNRs have cost advantage and financial advantage. The accession will bring MNRs more favorable conditions to develop this format nationally. Hypermarkets will be the main format that MNRs operate in China. The Chinese government is considering taking planning control in the store numbers and the store locations. Wal-Mart and Carrefour will be the main players in this format in the short term and the Chinese market could be an important battlefield to determinate their positions in global retailing.

Specialty store. This format has the highest profit margin among all retail formats in China. The Chinese specialty store traditionally focused on clothing but it is diversifying into furniture and home appliance electronics. Chinese specialty store operators have controlled the home appliance electronics market and will continue to dominate the field. Gome, Suning and Sanlian are the leaders in the format. Gome has grown the fastest in this format during the past three years. It planned to open 200 stores by 2003 targeting the sales of RMB 20 billion Yuan. Foreign operators IKEA, Q&D, Home Depot and Leroy Merlin are becoming the leading players in the Chinese furniture market. The accession will bring great opportunities in this format and more foreign entrants. The hyper-specialty store will be a trend. Foreign and Chinese retailers will share this market.

Discount store. This format had not emerged in China by 2002. A discount store normally purchases directly from producer, focuses on popular brands and locates at a low rent position. Dia under Carrefour will open the first discount store in China in 2003. Lianhua will launch this format in 2003, too. The format will focus on price sensitive products, particularly food and daily consumer products. Dia plans to open 300 stores in China by 2006. This format will be popular and directly compete with the supermarket and hypermarket. The accession will drive the emergence and the development of this format in China.

12.1.3 The impact of the accession on different regions

The impact of the accession varies in different regions in China due to the different structures of local retailing, different development levels of local economies and different consumption capacities. In cosmopolitan cities such as

Beijing and Shanghai, where are the most developed with great consumption capacities, retail giants, especially MNRs such as Wal-Mart and Carrefour, will mainly compete for these markets and may achieve obvious advantages there. In other major cities, with populations of about 4 million each, Chinese retailers are dominating the areas with department stores and chain supermarkets, and MNRs will target these regions for their next round of expansion. In the SM sized cities and rural areas, because of less economic activities and low consumption levels, it is Chinese local retailers that play the game there; chain operation can be a good choice and the impact of the accession on them will be the least.

The dual structure in the Chinese economy will also influence MNRs' expansion. They will stay mainly in Eastern and Southern China and large cities in other regions, while Chinese retailers will dominate the other regions particularly Chinese rural markets and SM sized cities. "Individual MNRs might do well, but the market will remain dominated by domestic retailers," said Mr. Huanghai, a senior officer in charge of Chinese retailing.

12.1.4 The different impacts of the accession on Chinese retailers and MNRs

To MNRs, China's accession to the WTO means more opportunities than challenges. They can develop the Chinese market with fewer restrictions. In the past decade, MNRs have been intimidated by Chinese restrictive regulations, chaotic distribution and fragmented retail channels, which will be solved gradually after the accession. The opening-up of Chinese logistics will boost the retailing development. The last winners will be those who can build and control their own complete distribution systems and successfully realize their localizations in China. For Chinese retailers, the accession means more challenges than opportunities. The accession is a serious challenge to Chinese retailers, but not all of them can completely understand this. After China's accession to the WTO, the direct result will be the fast increasing market share of MNRs. In large Chinese cities, MNRs may take as many as 40 per cent market share within the next five years. Many Chinese retailers, especially those who cannot successfully develop economies of scale and core competencies, will face the fate of going under. In fact, the strategy that the Chinese government has taken during the opening of Chinese retailing is to weaken MNRs' competency by separating their retail operation from other supporting industries, for example, foreign operation in the logistics and distribution services was not allowed in China, and by tightly controlling their chain operation, for example, chain operation is not allowed without getting permission from the State Council and Chinese companies must take the majority stakes in the JVs whose chain stores number more than 30. This "dismemberment" way makes most MNRs lose their competitive edge, which provides Chinese retailers with an opportunity to grow.

Therefore, Mr. Huanghai said that "Chinese retailing will be dominated by domestic retailers, even if MNRs are to take 20–30 per cent of the market share; local counterparts would still have room to develop." He argues that this was possible because of China's vast market potential and opportunities from the imbalanced economic development. When MNRs are developing the Chinese market, some Chinese retailers are also taking internationalization, such as Tian-ckro, who opened its first overseas hypermarket in Moscow in 1999 and has successfully opened three hypermarkets in Russia. It is exploring a new model in its internationalization, which is to integrate its retail business with its international trade. Actually, Tian-ckro is one of the most profitable Chinese retailers whose profit increase is always over its sales increase. In 2001, its sales increased by 34 per cent while its profit increased by 300 per cent. In future, there will be more Chinese retailers exploring international markets. With the further globalization of world retailing and China becoming the world "workshop," the time when Chinese retailers march into the global retail market will come sooner or later.

12.2 Chinese retail policies to meet the challenge of accession to the WTO

To meet the challenge of China's accession to the WTO, the Chinese government mainly takes two ways improving the competitiveness of its retailing: building the "national teams" and making new regulations.

12.2.1 Building large Chinese retailer teams by merger and acquisition

In the retail industry, size does matter. Compared with MNRs, Chinese retailers are much smaller and weaker. To be able to compete with foreign retail giants, Chinese retailers have to be strong enough. Since 2003, the Chinese government has begun to build 15 to 21 national teams, large Chinese retail groups, by restructuring the current state-owned retailers. A significant case is the formation of Shanghai Brilliance Group (SHBG) (Figure 12.2).

In 2003, supported by the Shanghai government, the Chinese retail giant, SHBG, was built by merging three large Chinese retailers: Shanghai Yibai Group, Shanghai Huanlian Group and Shanghai Friendship Group, and one Chinese

Figure 12.2 The structure of Shanghai Brilliance Group

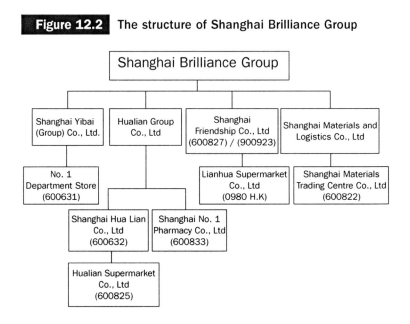

large logistics company, Shanghai Materials and Logistics Corporation Limited. The SHBG have 7 listed companies and include the largest Chinese department store and the largest Chinese supermarket chain. In 2003, Lianhua's sales were RMB 24 billion Yuan with 2,542 stores; Hualian's sales were RMB 18 billion Yuan with 1,325 stores; the sales of the Shanghai General Merchandise Shopping-Center (GMS) were RMB 3.5 billion Yuan with 20 stores and Lawson's' sales were RMB 350 million Yuan with 144 convenience stores.

The SHBG is the largest Chinese retailer with over RMB 33 billion Yuan in assets, over 200,000 employees and over 4,500 stores in more than 20 provinces. Its annual sales reach RMB 90 billion Yuan, which takes the company within the 100 world's largest retailers and the Fortune Top 500 Companies. It targets RMB 120 billion Yuan of annual sales by 2010. The Company plans to merge its 828 supermarkets, 11 hypermarkets and over 1,000 convenience stores under the Lianhua brand, expanding to 8,000 outlets nationwide by 2008. The core business of the company is hypermarket, supermarkets and convenience store. This emergence of the Chinese "Wal-Mart," the SHBG, will have a significant influence on the Chinese retail industry and particularly the competition situation in the industry.

12.2.2 Making new regulations and laws

To meet the challenge of China's accession to the WTO, the Chinese government is making some new regulations to further open the retailing in one hand while to regulate the industry in the other. On 31st January 2003, the Ministry of Commerce, which was a merger of the former MOFTEC and the National Bureau of Internal Trade at The SETC, issued the "Provisional Rule on the Establishment of Sino-Foreign Equity JV Foreign Trading Companies," by which foreign investors may build JV trading companies to import and export goods and technologies independently, and to engage in wholesaling of their import goods in China. There is no geographical restriction for the building these trading companies. This regulation makes it easier for MNRs to developing differentiation advantages in China by importing more foreign products and to developing cost advantages by exporting Chinese products and selling them in their international outlets. After accession to the WTO, the Chinese protective policies are gradually easing. MNRs have more choices in the Chinese market: they can make China either their procurement center or their market, focusing on store opening, or indeed, both of them; while Chinese retailers only have one choice: they have to grow fast and strong to be able to compete with foreign retail giants. MNRs can attack or withdraw in the Chinese market, while

Chinese retailers have only one choice: they have to compete in the market in order to survive.

In the transitional economic period, Chinese retail competition and the development of the Chinese retail industry often present a disordered picture due to the lack of experience of the Chinese government in managing the opening of the industry. For example, many local governments have opened their retail markets further than the WTO timetable and the Chinese government permit. Meanwhile, many MNRs broke Chinese regulations to open more stores than they should and have entered local markets where they are not allowed in order to achieve the FMA in the markets. So, the Chinese government has begun to strengthen state planning for national retail development since 2003. In early 2003, the Chinese government asked local governments to draw up detailed plans for retail development in their own regions. Those local governments which failed to submit their plans would not be allowed to approve new foreign invested retail projects. In addition, the Chinese government promotes a hearing system and encourages local governments to introduce hearing systems to approve the applications from foreign retail companies. These measures will make the development and opening of Chinese retail industry healthy and well under the government control.

12.3 The fast change in Chinese retailing

A significant and typical characteristic of Chinese retailing is its fast changes due to the Chinese transitional economy. In 2003, two important changes happened in the Chinese retail industry: forming the Brilliance Group and launching new retail regulations.

In the retail industry, size does matter. Compared with foreign retailers, Chinese retailers are much smaller and weaker. To be able to compete with foreign retail giants, Chinese retailers have to be strong enough, which normally can be achieved by mergers. Supported by the Shanghai government, the Chinese retail giant, Shanghai Brilliance Group (SHBLG), is merging with three large Chinese retailers: Shanghai Yibai Group, Shanghai Huanlian Group and Shanghai Friendship Group, and one Chinese large logistics company, Shanghai Resources Corporation. The SHBLG will have 7 listed companies and include the largest Chinese department store and the largest Chinese supermarket chain. The SHBLG will be the largest Chinese retailer with over RMB 33 billion Yuan in assets, over 200,000 employees and over 4,500 stores in more than 20 provinces. Its annual sales can reach RMB 100 billion Yuan, which takes the company within the 100 world's largest retailers and the Fortune Top 500 Companies. The core business of the company will be department stores, supermarkets and logistics services. This emergence of the Chinese "Wal-Mart," the SHBLG, will have a significant influence on the Chinese retail industry and particularly the competition situation in the industry.

After accession to the WTO, Chinese retailing is opening wider and wider, which brings more favorable conditions for foreign retailers. On 31st January 2003, the Ministry of Commerce, which was a merger of the former MOFTEC and the National Bureau of Internal Trade at The SETC, issued the "Provisional Rule on the Establishment of Sino-Foreign Equity JV Foreign Trading Companies," by which foreign investors may build JV trading companies to import and export goods and technologies independently, and to engage in wholesaling of their import goods in China. There is no geographical restriction for the building of these trading companies. This regulation makes it easier for foreign retailers to develop differentiation advantages in China by importing more foreign products and to develop cost advantages by exporting Chinese products and selling them in their international outlets. After accession to the WTO, the Chinese protective policies are gradually easing. Foreign retailers have more choices in the Chinese market: they can make China either their procurement center or their market, focusing on store opening, or indeed, both of them; while Chinese retailers only have one choice: they have to grow fast and strong to be able to compete with foreign retail giants. Foreign retailers can attack or withdraw in the Chinese market, while Chinese retailers have only one choice: they have to compete in the market in order to survive.

In the transitional economic period, Chinese retail competition and the development of the Chinese retail industry often present a disordered picture due to the lack of experience of the Chinese government in managing the opening of the industry. For example, many local governments have opened their retail markets further than the WTO timetable and the Chinese government permits. Meanwhile, many foreign retailers have broken Chinese regulations to open more stores than they should and have entered local markets where they are not allowed in order to achieve the FMA in the Chinese local markets. So, since 2003, the Chinese government has begun to strengthen State planning for national retail development.

In early 2003, the Chinese government asked local governments to draw up detailed plans for retail development in their own regions. Those local governments which failed to submit their plans would not be allowed to approve new foreign invested retail projects. In addition, the Chinese government promotes a hearing system and encourages local governments to introduce hearing systems to approve the applications from foreign retail companies. These measures will make the development and opening of Chinese retail industry healthy and well under the government control.

Although China presents tremendous opportunities for Chinese and MNRs, the fast changes in the Chinese transitional economy and Chinese retail industry ask both of them to keep learning and to keep examining their strategies so that they can change their operations according to Chinese market conditions, in order to be competitive. Otherwise, China will provide them with tremendous opportunities for failure. The future of Chinese retailing is uncertain, but it is full of promise.

12.4 Summary

Although China's accession to the WTO will bring a deep impact on Chinese retailing, the impacts vary greatly in different regions and in different retail formats, which provide Chinese retailers with opportunities to grow up and become competitive in some regions and some retail formats. Meanwhile, there is the three years' transitional period for Chinese retailers. Chinese retailers can take full advantage of this three-year period to prepare and to develop some advantages in some regions. Therefore, it is possible for Chinese retailers to be able to compete with MNRs and obtain advantages in the competition.

13

Conclusions

Can Chinese retailing compete? Is Chinese retailing different from other sectors and does it have more chances to catch up? There is enormous scepticism, from both inside China and abroad, about whether Chinese retailing will be competitive. The question should be answered by both the Chinese government and Chinese retailers together. Chinese retailing will be competitive if the Chinese government undertakes a systematic change and Chinese retailers apply new strategies. Actually, the answer is not one that can be chosen freely: Chinese retailing must be able to compete because of its strategic importance in Chinese transitional economy. Compared with other industries, Chinese retailing has more chances to catch up, because world retailing is still in its infancy in globalization and most MNRs still lack experience in internationalization, because imbalanced Chinese local economies and diversified local culture make it hard for MNRs to transfer their successful models developed abroad to the Chinese market and realize localization in the short-term; and because many Chinese retailers are growing fast, supported by the Chinese government.

13.1 The importance of Chinese retailing in the Chinese economy

Since China entered the buyer's market in 1998, competition among Chinese companies has become more and more intensive. After several years' competition, Chinese companies gradually understand one simple truth: in Chinese transitional economy, their competition actually is to compete for distribution channels, because in the buyer's market, retail stores have become a kind of scarce resource that the companies compete for. 2001 was called the year of "competing for distribution channels" by Chinese companies. It was the first time that they understood the importance of distribution channels in their competition, particularly the importance of the last outlet of the channels, retail stores, without which commodities cannot reach customers effectively. This is also the main reason that retailers can charge kinds of "channel fees" while suppliers have to pay them. In this buyer's market, the effect of Chinese retailing on Chinese consumption and up-stream industries is significant. It is not difficult to imagine what will happen to Chinese consumers, companies and the transitional economy if MNRs control Chinese retailing.

Competition for distribution channels indicates the importance of Chinese retailing in the whole production and the circulation system of commodities. Retailers normally have two essential functions: satisfying consumer demands and provision of outlets for production. The primary objective of reforming Chinese retailing is to generate favorable conditions for China's long-term economic growth. Reform of Chinese retailing will directly influence both Chinese consumption and production, by which influences Chinese economical growth.

According to the NBSC, in 2002, the contributions from investment, consumption and export to the growth of Chinese

GDP were 5: 4: 1. Mr. Li Rongrong, the former minister of SETC, said that the contribution of Chinese retailing to Chinese GDP was about 10 per cent in 2002. If it reaches 15 per cent as that of Western countries, Chinese economic growth will be greatly benefited. An unimpeded distribution channel not only reduces transaction costs and causes positive effects such as a "pull force" on production companies but also strongly stimulates domestic consumption by improving consumer welfare and meeting customer demands, such as easier access to cheaper and wider selections of products, which would contribute to Chinese economic growth.

Since its accession to the WTO, China has become the world's workshop. But if Chinese companies cannot control their distribution channels, particularly Chinese retailing, then both Chinese production companies and consumers may strongly depend on MNRs, which indicates that Chinese production companies will lose their independence and put their fate in the hands of foreign companies. Further, favorable Chinese retailing would promote Chinese products abroad through exporting Chinese products from both Chinese and MNRs. Some Chinese retailers such as Tian-ckro have opened supermarkets abroad specially selling Chinese products. In 2001, exports of MNRs accounted for 12 per cent of China's national export. In 2002, Wal-Mart's procurements from China accounted for 10 per cent of American imports from China. By 2002, although China was the fifth largest exporter in the world, China had been the country that was involved the most anti-dumping cases in the world for the past seven years. If MNRs control the Chinese market, the dominant power in Chinese retailing will be in the hands of MNRs; foreign products could reach Chinese consumers easily on a large scale and in vast quantities, because MNRs have powerful global procurement networks, which may harm both Chinese production and consumption by separating Chinese production companies from retailers and by controlling Chinese distribution channels. This situation will be very unfavorable for the developing Chinese economy. The importance of Chinese retailing also stems from its contribution to national employment. According to Mr. Li Rongrong, the total employment in Chinese retailing reached 47 million in 2002, the second largest employment in the non-agricultural sector and over a quarter of the third industry; retailing could create more job opportunities in the long run if the industry continues to develop at a high growth rate. In addition, if the reform in Chinese retailing succeeds, it could drive reforming Chinese up-stream industries and greatly affect production companies. Shultz (1993) points out that the future differentiation CA of a company comes from "channels" and "communications." As the final outlet connecting consumer, retailers influence production companies by being both the physical channel and the market information passing channel. Market oriented operation and fair competition encouragement in retailing will drive production companies to operate according to the laws of the free market. Thus the acceleration of institutional change from a transitional to a market economy will be generated.

Successful reform in Chinese retailing also brings political contributions to the Chinese government. There is political importance for Chinese government, which manages 1.3 billion consumers, 785 million of whom spend less than US$1 per day, in preserving retail price stability, transparency and ensuring competitive benefits are passed to the consumer. Further, developed and competitive retailing helps to keep inflation down. The contribution of Wal-Mart to keeping low inflation during America's high economic growth period in the 1990s has been noticed and discussed recently. Koretz (2002), the economist at UBS Warburg, points out that "the discounter drives prices down and that one reason the Federal Reserve is less concerned about inflation than the European Central Bank is the deflationary impact of America's more competitive retail environment." To China, keeping low inflation is very important when its economy grows at high rates in the next decade, because it helps to keep macroeconomic stability, which is important to the transition economy (de Larosiere, 2001). Chinese retailing must thus be able to compete. To do so, however, both the Chinese government and Chinese retailers must undertake some systemic changes.

13.2 The role of Chinese government in the retail change

China's accession to the WTO has meant that Chinese retailers cannot evolve naturally as western retailers did in their developing period. Chinese retailers have to compete with powerful MNRs and face serious competition conditions when they are very weak in their infancy. To improve the competitiveness of both Chinese retailers and retail industry, the Chinese government must undertake the responsibility to support developing Chinese retailing. Without the support of government, it is impossible for Chinese retailing to become competitive. From Porter's Diamond model, it may be seen that the Chinese government could play an important role in improving competitiveness of retailing by changing the unfavorable factors to favorable conditions. For example, in input factor conditions, the Chinese government could make policies to increase investment in retail research, retail personnel training and the application of new technology to retailing; in demand conditions, it could encourage Chinese consumers to increase consumption (Chinese RMB saving had reached RMB 10.4 trillion Yuan by 2003); in related industries, it could largely develop the logistics industry and increase investment in related public infrastructure; and in local rivalry, it could promote fair competition. The Chinese government should undertake at least the following main responsibilities in retail change:

1. It should play a critical role not only in driving reform but also in forming the rules of retail competition and in supervising the market orders. Competition is essential for a market to function efficiently; and it forces firms to look for more efficient ways of producing goods and to meet the desires of consumers more effectively. Government must set and enforce the basic laws of society and provide a framework within which firms can complete fairly against one another (Stiglitz and Driffill, 2000). Because China is in both economic transition and an industrial restructuring period, this function is particularly important. The way that the Chinese government supervises retail competition will determinate the rule of Chinese retailing.

2. It must pay more attention to promoting fair competition because unfair competition happens more frequently in the Chinese transitional economy than in a matured, free market economy. To avoid vicious competition and excessive wasting of resources, the Chinese government should introduce related policies to restrict the increasing disorder and unfair competition in current Chinese retailing and plan the future development of the retailing. Beaujeu-Garnier and Delobez (1979) show that the process of retail planning in the UK has resulted in different retail patterns from those in France and USA. The Chinese government also needs to make regulations in city planning for retail format control. Regulations restricting excessive competition are necessary. In promoting fair competition, the Chinese government needs to consider SM sized retailers; such as it needs to help or encourage them to build procurement associations and assist them to improve

their competencies by providing them with financial support for their development.

3. North (1990) points out that institutions affect the performance of an economy according to their effect on the cost of exchange and production. In Chinese retailing, transaction cost is expensive because the old institutions have been broken while new market-oriented institutions have not been yet established. Local protectionism, undeveloped infrastructure, bureaucratic administration and serious information asymmetry make transaction costs very expensive. Insufficient competition discourages many retailers from actively improving their technology and management to increase productivity, while mainly depending on charging of fees for their profits. It is the government's responsibility to reduce transaction costs and promote retail competition through reform, e.g. by regulating retailers in their freely charging fees could promote competition and drive retailers to improve efficiency.

4. The Chinese government also needs to promote applications of technology. The wide application of new technology contributes to retail development and improvement of competitiveness.

5. It must promote entrepreneurship throughout the society and encourage the building of incentive systems in Chinese retail companies. One main reason for there has been few successful retailers in China is that there are few successful entrepreneurs. The Chinese government should further reform SORs and particularly its reward system, which is an important way to build an incentive mechanism and promote entrepreneurship in the nation.

6. To form a united market by lifting local protectionism, strict regulations and policies to eliminate local protectionism are needed. This concerns the government's ability

to formulate right industrial policy and implement industrial policy. Since the Chinese government has begun to reform the retail industry, its reform direction is right, but it ignores reform of those industries that support retail reform and coordination between retailing and its related industries. Reforms in logistics industry and in promoting entrepreneurship lag far behind. Meanwhile, in the transitional economy, the government's ability to implement industrial policy is also important. The local approved JV phenomenon and the obstacles in developing chain operations in China show that the ability of the Chinese government to implement its industrial policy is weak. The Chinese government must improve this ability, without which, developing competitive Chinese retailing remains difficult.

The dominant role of the Chinese government in the Chinese economy and industrial development will remain, but the nature of that role must be fundamentally altered. It must change from protecting, cushioning and sheltering Chinese retailers to create a fair and rational environment for retail competition and greatly encourage fair competition. However, change in government policy and function alone is not enough to improve industrial competitiveness; Chinese retailers themselves must also undertake fundamental changes to meet the challenge of accession to the WTO. They must develop a new corporate model and compete in strategy.

13.3 Transforming Chinese retail companies

The homogeneous phenomenon in Chinese retailing indicates that few Chinese retailers have strategies. A successful retail strategy requires real innovation and making trade-offs among retailers, suppliers and customers. China's accession to the WTO presents serious challenges to Chinese retailers. To meet the challenges, Chinese retailers must develop their own strategies, which could include at least the following.

Growth strategy. Canals (2000) argues that corporate growth is important because it is critical in managing revenue and costs, developing talent, attracting capital and breaking the mature-industry mindset. But to Chinese retailers in the transitional economy, there is another important reason, which is that size, scale and strength are essential to surviving increasingly intensive competition in Chinese retailing, given the great strength gaps between Chinese and MNRs. For example, Wal-Mart's annual sales were over 130 times of that of Lianhua in 2002. Even the sales of the 100th largest global retailer were over three times those of Lianhua. With such great difference, how are Chinese retailers able to compete with MNRs once Chinese protection policies are all lifted after December 2004? Chinese retailers must try their best to grow up first in order to compete with MNRs. Further, fast growth could make Chinese retailers easily recognized by capital markets and acquire good funding ability, which is important in retail competition because most Chinese retailers lack enough capital funding for their expansion. Corporate growth normally takes three forms: (1) concentrating on a single business in its home country, (2) seeking vertical integration, either upwards or downwards, and (3) diversifying current business or expanding to other countries. Currently, Chinese retailers need to concentrate on their retail business domestically; and the key question for them is how to grow larger and faster. Normally there is a choice in corporate growth: organic growth or by M&A. Organic growth has many advantages, e.g. it is less risky; capabilities could be well developed. However, organic growth takes time. Further, there is an important obstacle preventing organic growth in Chinese retailing: appropriate locations for store opening are scarce. In addition, an aggressive influx of MNRs has squashed the average gross profit margin of the industry. Chinese supermarket's gross profit margin had been reduced from 20 per cent in the mid 1990s to about 8 per cent by 2000, which is very unfavorable for Chinese retail organic growth. In large cities such as Beijing, cut-throat competition had already reduced retail profit margins to 3.7 per cent in 2001. The fundamental solution to the growth problem is scale expansion and institutional streamlining. In world retailing, restructure and reorganization have become the trend since the 1990s. Given the fragmented and low profit margin of Chinese retailing, consolidation by M&A could be the best choice for Chinese retailers' corporate growth. Some Chinese retailers had taken consolidation before China's accession to the WTO. Beijing supermarket chains Chaoshifa and Tianckro announced their merger just one week after China joined the WTO and became the 15th largest chain retailer. Local governments also promote consolidation. The Beijing municipal government, e.g., is striving to consolidate the city's more than 3,000 retail companies to form five large supermarket groups. However, there have been few successful cases in the M&A yet. Chinese retailers must pay attention to some problems in their M&As. One is integration after M&A. M&A is just the first step in consolidation and does not guarantee success. International M&A experience has shown that many M&A cases fail due to the failure of integration. Therefore, Chinese retailers should focus on the integration of corporate resources after their M&As. Another problem is to achieve balance between rate of growing and improvement in management. An important feature of Chinese retailing is that many enterprises grow too fast. In the retail industry, size clearly does matter. But Chinese SOEs have proven in the past that size is not everything. Often their most pressing problem is poor management and abysmal service. "Retailers, especially after merging, need to improve the standard of their service," says Zhang Wenzhong, the CEO of Wu-Mart, one of the largest Chinese retailers. Chinese retailers must promote improvement in management by investing in infrastructure including building ERP system and wide application of IT. In addition, there is a problem of funding corporate growth. Listed in stock markets and then taking capital operation is highly recommended for its lower capital cost than other ways. Although growth strategy is important for Chinese retailers, they must solve the kinds of problems emerging in their growth in time and pay attention to growth quality rather than just increasing their store numbers.

Alliance strategy. Chinese retailers often called MNRs "wolves" and shouted "wolves are really coming" when China entered the WTO. Facing these "wolves," Chinese retailers may have three options: one is to become "wolves" and dance with them; another is to become "wolves' friends" and cooperate with them; and the last option is to become "sheep" and be eaten by them. Some Chinese retailers are trying to be "wolves" by growing stronger. But for some Chinese retailers, being "wolves' friends" is not a bad choice, because some MNRs need such cooperation for several reasons. Firstly, because of regulation, within the 3 years since the accession, although China has lifted nearly all restrictions on retailing, Chinese retailers must retain a majority share in the retail JV with chain store numbers over 30, which provides foreign and Chinese retailers with an opportunity for cooperation. Another reason is that some MNRs need local retailers in their expansion and localization. Some MNRs may ask for cooperation to obtain special resources controlled by local retailers while MNRs cannot acquire them in other ways, such as good locations for store opening and establishing distribution networks. Other MNRs may want to realise localization by cooperating with local retailers. Through cooperation, both foreign and Chinese retailers can benefit each other, and it may be a win-win strategy; but to develop an alliance strategy, Chinese retailers must develop firm specific advantages first.

Alliance can also be made between Chinese retailers. In order to defend competition from MNRs especially Wal-Mart and Carrefour, any Chinese retailer alone is not able to compete with them. But if several Chinese retailers ally together, they may be more competitive in a local market. Meanwhile, MNRs are stronger than their Chinese counterparts; but when they compete throughout the whole nation their strength may be weakened for their resources are divided into many regional markets. Such alliance could be in many fields, such as in sharing distribution systems, cooperating for procurement, controlling or monopolising of sales of popular products. Alliance strategy can also be used in store expansion. In November 2001, seven Chinese local retailers and six Chinese investment companies announced a plan to pool financial resources together to open 1,000 stores in the Beijing area within the next five years. Besides, in Beijing, some local retailers developed several flagship supermarket groups to enhance their competitiveness. In March 2001, the Beijing Chaoshifa Company, Xidan Department Store and Hualian began business and aimed to open 1,000 outlets within next the three years. Although Chinese retailers are weaker, some of them can cooperate together in regional markets or focus on one regional market to compete with MNRs, which may make Chinese retailers have more chance to win MNRs.

Right positioning and developing CA strategy. Positioning identifies a retailer's target market while developing CA provides the retailer with a powerful means to reach the market. A retailer's positioning also indicates its competitors; and different positioning strategy brings the retailer different competitors. Only if Chinese retailers take appropriate positioning first, can they then develop their own CAs. A general will lose his battle if he cannot correctly judge who his enemies are. A main reason for many Chinese retailers' failure is their wrong positioning; the vicious competition focusing on price war between Chinese department stores and supermarkets is an example in this regard, for they compete for the same segment. Facing foreign competition, some Chinese retailers may target niche markets by developing their unique CAs, such as personalized service focusing on a certain small segment. Such CAs could be developed by establishing partnerships with other companies, either horizontal to convey added value by alternative distribution channels or vertical to achieve cost leadership at various stages along the value chain by supply chain integration, cooperation in logistics and inter-organizational process optimization.

After right positioning, Chinese retailers should try their best to develop their CAs by exploring their FSAs and taking advantage of China's CSAs. In addition, Chinese retailers should develop their CAs through innovation. Just copying what MNRs do, or only improving one aspect of retail operation such as the quality of goods, is not enough for Chinese retailers to win the competition with foreign retail giants. Peterson and McGee's (2000) studies suggest that sustained marketing emphasis on quality is not a clear prescription for removing the impact of Wal-Mart presence. The case of Vanguard is a good example of the importance of innovation in developing CAs. Among kinds of CAs, widely recognized brand reputation is essential. Chinese retailers need to build a strong store brand image in order to compete with MNRs. To do so, besides appropriate promotion, they also need to resort to modern technology to upgrade their operations. Excellent streamlined logistics is critical, because it not only guarantees high product availability but also improves profit margin by reducing cost. On the other hand, to develop CAs, Chinese retailers also need to overcome their weaknesses. In Chinese retailing, a common characteristic among Chinese retailers is their high debt rates and low profit margins. Hualian, one of the best Chinese retailers, had just about 3 per cent profit margin while its debt rate in 1999 was as high as 77.38 per cent; thus, its capability for future expansion is doubted (*South Daily* 03/05/2000). This weakness is fatal to Chinese retailing. Likewise, another fatal weakness is their relations with their suppliers. Good supplier relations are the source of all CAs. Great dependence on charging kinds of fees for profit and poor credit in suppliers' payments not only impairs bilateral relations but also makes a retailer's CAs lose their roots. So developing a multi-benefit relation between retailers and suppliers is critical for Chinese retailers to succeed.

Focus strategy. During competition with MNRs, focus strategy is particularly important for Chinese retailers. Because single Chinese retailer's resources and capabilities are very limited, centralized use of limited resources, such as concentrating its corporate resources on a certain market or a specific retail format, could bring the retailer more possibilities or better chances of success. Focusing on the market where the retailer has greater local advantage may make the retailer more likely to succeed. In 2000, the two largest Chinese retailers, Lianhua and Hualian claimed that they would cooperate in opening 1,000 stores in the Eastern, Southern, Northern and Middle China to be a "Chinese Wal-Mart". However, Hualian's 2001 annual report showed its main sales still came from its home market, Eastern China, and its new business was less than 3 per cent of the total. It may be argued that Hualian should adopt a focus strategy rather than ambitiously compete too widely. Similarly, many Chinese retailers' first reaction to China's accession to the WTO is to compete for outlets; they are trying their best to take over as many locations as possible without considering their capabilities and resources. Then limited resources and capabilities cannot support the newly opened stores and retailers have difficulty in developing their CAs. Focusing on the single format where the retailer has the greatest expertise and advantage makes the retailer more likely to succeed. Generally speaking, developing the right strategies is critically important for Chinese retailers' success. However, one strategy alone may be not enough for Chinese retailers to succeed. They may pursue several different strategies according to their own specific conditions; e.g. dual strategies, which combines growth strategy and developing CAs strategy. Developing the right strategies can make Chinese retailers change from competing for one or several aspects of the retail process to compete for the whole supply chain and successfully achieve both inward focus and outward focus. Meanwhile, Chinese retailers need focus on leveraging local assets in market segments where multinationals are weak and apply defensive strategy as Dawar and Frost (1999) recommend. When Chinese retailers are armed with distinctive strategies, as with Vanguard in Shenzhen, they can be highly competitive and profitable during the competition with foreign retail giants.

13.4 Strategies for Chinese retailers to compete with MNRs during the WTO membership times

To compete with foreign retail giants, Chinese retailers must develop appropriate strategies. Strategy is vital, because it guides the direction of corporate competition. Although

MNRs are very strong, they have weaknesses; Chinese retailers should exploit and take advantage of these weaknesses. To different Chinese retailers, their advantages and the competitors may be different, so their strategies should not be the same. Chinese retailers might take the following strategies to compete with MNRs.

1. **Focus and saturation strategy**. When Chinese retailers develop their strategies, the first issue that they must consider is in which market to compete: in which geographic market they should compete in and which customer segment they should target. No current Chinese retailer is able to compete in the whole nation, so they need to focus on regional markets, particularly their home markets where they have strong local advantage. When choosing a market in which to compete, Chinese retailers should thus determine where they have greater advantage than MNRs, or where MNRs are weaker. Some strong and large Chinese retailers should focus on urban markets, while weak and small retailers are better advised to avoid direct competition with foreign retail giants and take time to grow first. The suburban and rural market, where they can succeed more easily for less intensive competition, could be a good choice. Choosing targeting segment is to choose retail format. Because Chinese consumption is still low, the mass market is a good choice; and accordingly the retail formats chosen could be either hypermarket or supermarket. They should then adopt a focus and saturation strategy, which is to focus on a single market, their home or nearby market, and concentrating on operating one single retail format. They should not expand to other markets unless they have dominated the market as did Wal-Mart in the 1970s. In order to secure market saturation, Chinese retailers can either use a single retail format, or develop multiple formats. Normally the former should be priority and multiple formats be developed only after they

have succeeded in the single format. When they have dominated the home market and expanded to neighbor markets, the single format should normally be the first choice; using multiple formats to develop several markets is normally not encouraged due to high risk (Figure 13.1). My research indicates that nearly all Chinese retail giants have taken expansion strategy for developing new markets in other provinces; however, few of them have succeeded. Chinese retailers should take a focus and saturation strategy. These days, MNRs only compete for large cities. Chinese retailers should try to avoid direct competition with them at the current stage and focus on markets where competition is less intensive such as suburban markets, rural markets and SM sized cities as Suguo has done. Some large Chinese retailers may choose cities where MNRs have not entered or competition is less intensive for expansion. Lianhua, for example, is exploring Central and Western China; it opened stores in Xi'an and Chengdu in 2000.

2. **Gradual and flexible expansion strategy.** This strategy means that, during its expansion, a retailer should focus on its home market and then gradually expand to its neighbor and further cities by taking different operation methods according

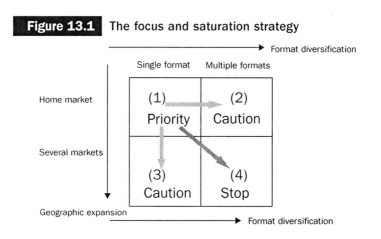

Figure 13.1 The focus and saturation strategy

257

to local conditions. It thus easily develops a complete supply chain system. Establishing a home base is essential to support a retailer's further expansion. Without a solid base market, the expansion is like a tree without roots and a river without a source, and the expansion cannot last long. This strategy is low risk and economies of scale are easily achieved. Lianhua, for example, has made Shanghai its base and gradually expanded to its neighbors, such as Zhejiang Province and Jiangsu Province, by different ways adapting to local conditions: direct investment for hypermarkets, building JV for hypermarkets, merger and acquisition for standard supermarkets, franchise for developing standard supermarkets and convenience stores. A flexible strategy is important because of the diverse conditions of markets. By choosing different ways to develop these markets, a retailer can fully take advantage of local resources and thus easily succeed.

3. **Differentiation strategy.** As mentioned above, Chinese retailers should exploit MNRs' weaknesses. One way to do so is to develop a differentiation strategy. Besides the weakness of geographic choice, MNRs have another weakness in product assortment, e.g. in PL product. By 2001, few MNRs had PL products in their Chinese operation and nearly all their products were sourced locally with local brands. PL product is an effective tool in building customer loyalty and developing differentiation advantage if it is used appropriately. This could be achieved by either focusing on low price with reasonable quality, e.g. Tesco's PL products, or by focusing on high quality with high price, e.g. St. Michael brand products of Marks & Spencer. While developing PL products, a retailer needs an effective control of its supply chain such as delivery reliability and product design. Since 2002, some MNRs have begun to consider developing PL products in their Chinese stores. In the past

two decades, Coca-Cola and Pepsi have dominated the Chinese carbonated beverage market, but the situation may be changed by foreign retail giants. Wal-Mart is cooperating with Cott, the world's largest private brand cola producer, to produce cola particularly for Wal-Mart stores in China. Cott has built a JV in China, which is called Sanhe Meile Limited Co., According to its CEO, Mr. Dai Jun, the negotiation with Wal-Mart is going well. Carrefour is also negotiating with Danone to produce Cola for Carrefour stores in China. Carrefour is planning to develop its PL products in two ways: firstly by transferring its global PL products to China; secondly by developing new PL products specially designed for the Chinese market. Chinese retailers should also pay attention to developing this kind of differentiation strategy. Lianhua was the first Chinese retailer to develop its PL products and has achieved great success in developing its differentiation advantage. Another way to develop differentiation strategy is repositioning its retail formats in competition. That many Chinese department stores and supermarkets compete for the same consumer segment indicates the competition between them is intense and both need to develop differentiation strategies.

4. **Winning in the transitional period strategy**. This strategy refers to Chinese retailers should seek to develop as many advantages as possible over MNRs in the transitional period so that they can achieve favorable conditions for further competition. The transitional period has two meanings: one refers to the period during which MNRs have not adapted to the Chinese retail market. Retail is detail and retail is local. Retailing is strongly influenced by local culture. Even Wal-Mart and Carrefour, who have been in China for over 7 years, still need time to adapt to local markets and realize localization. The other refers to the three years' transitional period in China's accession agreement. After China's

accession, there is a three-year transitional period from December 2001 to December 2004, during which MNRs are still restricted by Chinese policies from further development in China. Taking the opportunities that MNRs lack completely to adapt to Chinese markets and Chinese retailing policies still impose some restrictions on MNRs while protecting Chinese retailers, Chinese retailers should try their best to develop quickly in order to establish favorable bases for future competition. Chinese retailers must take advantage of the transitional period and try to achieve as many as advantages over MNRs in the period. Otherwise, Chinese retailers will find it more difficult to beat MNRs after the transitional period. The old saying is that a good beginning is half of success. Chinese retailers must focus on the competition at this stage. This means within the first three years of China's accession to the WTO, Chinese retailers must successfully (1) develop competitive operation models in their target markets and segments; (2) establish their base areas; (3) grow strongly enough with complete distribution systems and (4) develop their right corporate strategies for further growth.

5. **Capital operation and low cost expansion strategy.** A main obstacle preventing Chinese retailers from expanding is their lack of capital, while organic growth is slow and full of risk in current conditions of intensive competition. Many Chinese retailers may not have the chance to grow if they only pursue organic growth because powerful MNRs may put them out of business when they are weak and small. To solve the problem of capital, Chinese retailers can normally either borrow from banks or be listed in stock markets. Borrowing from a bank has two disadvantages: the amount of money borrowed is limited and the loan period is short. For most Chinese retailers with normally 70 per cent debt ratio, this is not a good choice: sourcing capital from capital

markets is a better choice. Chinese retailers should try to be listed in stock markets for borrowing enough capital for their expansion. Capital operation means a Chinese retailer, supported by capital markets, either merges with other retailers, or acquires other retailers, or controls other retailers by becoming their main shareholder; then listing its controlled retailers in stock markets, and using the money obtained from stock markets to control other retail companies. In this way, the retailer can grow quickly and expand fast. Those retailers that cannot be listed should either attract external investors by changing their corporate ownership structure, or ally themselves with other retailers, or be merged or acquired by other retailers, or to be a niche player. For these retailers, making an alliance, such as building procurement alliance, may be a good choice. In addition, SMRs can ally for procurement by establishing a company. On 23rd September 2002, nine retailers invested together to build the company, Shanghai Zhongyong Tongtai Electrical Appliances Marketing Ltd., for sourcing electrical household appliances.

6. **All-round competition strategy**. When Chinese retailers compete with MNRs, competition should focus on the whole retail system rather than one aspect of it, such as store operation. The iron chain tells us that the weakest link of the chain determinates its strength. Similarly, in competition, the weakest aspect of a retailer gives the limit of its competency. Focusing on just one aspect of the retail process and ignoring the whole operation system makes the retailer vulnerable to competition. With Chinese retailing opening, competition is becoming increasingly fierce and a price war happens frequently. In the past 5 years, many small retailers have been bankrupted while many large retailers have consolidated. In Shanghai, there were 28 supermarket chains in early 1990s while just 10 were left by August 1999. Recently, some

domestic chain stores have closed down one after another, while Carrefour, Wal-Mart and other foreign retail stores are seeing flourishing business in China. Competition has changed from competing for goods in the past to competing for the entire retail process: commodities, service, communication and distribution. Retail competition is now for the whole system rather than one or two aspects of the retail process. Chinese retailers have much to do to catch up with their foreign competitors. They urgently need to upgrade their management skills and build business credit to compete with foreign retail giants to shake up their managerial systems. However, most Chinese retailers still depend on the "channel fee" to survive, which accounts for over 70 per cent of profit for many large retailers. Many retailers choose suppliers by payment of the channel fee rather than the goods supplied, which often ignores customer demands. In the long run, a store that ignores customer demands and does not focus on developing its CA will fail in intensive competition. Many stores shout, "The Customer is God" while they do not fully consider God's interests and study God's demands. In order to compete with MNRs, Chinese retailers should improve the competency of the whole retail system from supply chain to CRM, particularly to overcome the weakest part of their retail system (Figure 13.2). In China, the successful retail model normally includes three aspects: the supply chain, corporate operation and CRM, while most Chinese retailers are weaker than MNRs in these three respects. Chinese retailers should thus develop CA at least in one of the three aspects. To develop an effective supply chain, especially an efficient distribution system, establishing partnerships with suppliers is essential; to establish, efficient CRM, the Chinese retailer must systematically study local customer demands and buying behavior; and to improve store operation, the Chinese retailers must update their management know-how.

Figure 13.2 Competing the whole retail system

Further, from Chinese retailers' cost structure, it can be seen that Chinese retailers' main profits are also from the three aspects (Figure 9.3): (1) Cost control in procurement. Procurement cost normally accounts for about 83 per cent of the total cost. Large quantity procurement can achieve cost advantages through strong bargaining power, and profit can be made from fast-moving products by high volume sales. (2) Logistic cost accounts for about 5 per cent of cost. Reducing cost from distribution system by improving efficiency can increase profit. (3) Fresh food and PL products, which can build customer loyalty, and PL products can normally generate a higher profit margin. But to achieve these, a retailer must understand consumer demands well, and this involves supply chain management, CRM and store operation.

So for Chinese retailers, it is not wise only to fight price war with foreign retail giants. Low price alone cannot meet customer demands and could not be a CA; a price war without support from a complete supply chain is short sighted and suicidal; service and shopping environment are becoming more important. When customers shop, they not only buy a commodity but also a "package," which includes service and shopping environments. My research in Shenzhen finds that price issues lie just behind "service" and "shopping environment," ranking it the third important issue that concerns customers. In intensive competition where all retailers are similar in their product assortment and prices, developing appropriate service could be a CA. However, few

Chinese retailers have developed service advantages. In future competition with MNRs, Chinese retailers must develop their own service model by exploiting their local advantages; and those who are good at developing service advantage will be more likely to succeed.

7. **Chain operation and branding strategy.** To succeed, Chinese retailers must develop chain operation on the one hand while emphasizing brand building on the other. Chain operation, either by direct investment or by franchising, makes retailers develop economies of scale and expand fast. A strong brand is easily recognized by customers and necessary to chain operation. Brand building is very important, because when a company creates strong brands, it attracts customer preference and builds a defensive wall against competition. But many Chinese retailers are unwilling to invest in building brands. Worse, they often ignore long-term brand building and reduce investment in brand support during their fast store expansion. Establishing a strong brand requires long-term focus and investment. Chinese retailers must deal with the relationships between brand building and chain operation well. Without a strong brand, chain operation cannot be effective.

13.5 Principles for Chinese retailers in their competition

Besides the above strategies, Chinese retailers also need some principles of operation in competition:

1. Take their home markets first, take small and medium sized markets and extensive rural areas in priority and take big cities later. 2. Do not rashly launch price war alone; launch no competition unprepared; launch no battle that is not sure of winning; make every effort to be well prepared

for each round of competition, particularly in seasonal and festival competition. 3. Strive to wipe out competitors when they are in transitional period. At the same time, keep paying attention to competitors' weaknesses and launch competition against them. 4. Compete against weaker and dispersed retailers first; against strong retailers later. 5. Strive to the utmost to preserve own strength and destroy that of the competitor, which means exploiting own advantages to compete the disadvantages of competitors and concentrate absolute resource advantages to focus on the target market. 6. During expansion, capital operation and store operation like a man's right and left hands; they should be used both and cooperate well. 7. Build an incentive mechanism and keep it improving. The richest source of power to wage competition lies in the resources a retailer has and how they are organized. An incentive mechanism must be built so that the resourses are always in an efficient organized state.

China's accession to the WTO brings Chinese retailing both opportunities and challenges. There are those who question whether Chinese retailers can compete with foreign retail giants after the accession. Even some Chinese retailers themselves doubt this. This research gives an answer: Chinese retailing is able to compete and Chinese retailers can be competitive if the Chinese government undertakes a systematic change in retail policies, particularly pursues competition oriented reform and if Chinese retailers develop competition oriented strategies. The successes of Vanguard and Renrenle in Shenzhen have given confidence. Meanwhile, the accession not only indicates challenges to Chinese retailers but also to foreign retail giants. Although both Carrefour and Wal-Mart have been in China for over 7 years, they have not yet developed profitable models nor realised localization. In local competition, Chinese retailers are not as weak as MNRs expected. Chinese and MNRs have their respective advantages and disadvantages. Chinese retailers can win the competition with MNRs if they fully exploit own advantages, e.g. performing their innovation ability well and adopting competition oriented retail strategy as Vanguard did. As the Vice Director of Marketing at Wal-Mart, Mr. McMillan, said some Chinese retailers grew fast during competition with Wal-Mart by their quick learning; at first, they learned a lot from Wal-Mart; now Wal-Mart learns more from Chinese retailers. After China's accession to the WTO, it is time for Chinese retailers to embrace a new competition strategy and model based on deep understanding its strengths and weaknesses in their past operation. A successful Chinese retailer will not be a clone of American or European retailers, but a new and distinctly Chinese conception of retailer. The biggest enemy for Chinese retailers is Chinese retailers themselves rather than foreign retail giants. As Nolan (2001) points out, it is conceivable that China's industrial policies might be given fresh life with a renewed focus and sense of urgency due to the impending shock of joining the WTO under the terms agreed. Despite the enormous challenges presented by the global business revolution, it is possible to imagine a strategy that might lead to the growth of competitive large firms based in China. Having joined the WTO, Chinese retailing can take the opportunity to catch up, and the Chinese government must undertake some fundamental changes in restructuring the industry and its role in the industry's development. This is the only option, because China could not afford the failure of the industry. The fate of Chinese retailing is in the hands of Chinese retailers and the Chinese government.

Appendix A:
Some conditions for retail JVs

1. The requirements/or qualified for foreign partners in retail and wholesale JVs:

   - In retail, average sales turnover of at least US$2 billion in the three years prior to the application, and assets valued at US$200 million or more.

   - In wholesales, average annual turnover of at least US$ 2.5 billion over three years, and assets valued at US$300 million or more.

2. The requirements for the Chinese partners in retail and wholesale JVs:

   - Minimum RMB 50m Yuan in assets (RMB 30 million Yuan in the central and western areas).

   - Average sales turnover of RMB 300 million Yuan over the past three years (RMB 200 million Yuan inland).

   - If the local partner is a foreign trade enterprise, it must have a trading volume of at least US$50 million (including export worth US$30 million).

3. Minimum registered capital:

   - Retail JV: RMB 50 millon Yuan (RMB 30 million Yuan in central and western areas).

   - Wholesale JVs: RMB 80 million Yuan (RMB 60 million in central and western areas).

4. Other key points:

- Fees paid to foreign partners in retail and wholesale JVs for intangibles such as trademarks, trade name licensing and technology transfer cannot exceed 0.3 per cent of sales turnover each year, and cannot last more than 10 years.

- Franchising operations of any form not allowed.

- Imports may not exceed 30 per cent of sales turnover.

5. The provision also asked that the business may take the form of equity JV or cooperative JV but the wholly foreign owned enterprise was not allowed. The business scope is limited to the retailing of general merchandise and the commercial business of import and export. It is not permitted to engage in wholesale business or act as import or export agency for another company. The application procedure is: the details of proposed JV should be submitted to the State Council by the regional government for approval. The qualification of the parties to the JV should be examined by Ministry of Internal Trade.

Source: "Official Written Reply to the Problem of Using Foreign Investment in Commercial Retail Field" the State Council, July 1992

Appendix B:
Top 10 Chinese chain store operators in 2005

Table B.1. Top 10 Chinese chain store operators in 2005

Ranking	Retailer	Sales in 2005 (Billion Yuan)	Number of stores
1	Brilliance Group Co., Ltd	72	6345
2	Beijing Gome Electrical Holdings Lt.	50	426
3	Suning Appliance Co., Ltd	40	363
4	Vanguard	31.3	2133
5	Dashang Group	30.1	130
6	Beijing Hualian Group Holdings Ltd.	20.8	74
7	Wumart Group Holdings Group Ltd.	19.1	659
8	NGS Supermarket Group Ltd	17.5	1572
9	Carrefour	17.4	78
10	Shanghai RIMart Ltd	16	60

Bibliography

Barth, K., Karch, N. J., McLaughlin, K. and Smith Shi, C. (1996) 'Global Retailing: Tempting Trouble?' *The McKinsey Quarterly* 1: 118–19.

Beaujeu-Garnier, J. and A. Delobez (1979) *Geography of Marketing*, London: Longman Group Limited.

Blois K. J. (1989) 'Supermarkets and their role in Chinese retailing', *European Journal of Marketing* 23: 7–16.

Canals, J. (2000) *Managing Corporate Growth*. Oxford: Oxford University Press.

Dawar, N. and Frost T. (1999) 'Competing with giants: Survival strategies for local companies in emerging markets', *Harvard Business Review* (March/April): 119–29.

Dawson, J. A. (1993) *The Internationalisation of Retailing*, London: UCL Press.

De Larosiere, J. (2001) *Transition Economies*. Cambridge, MA: MIT Press.

Hoch, S. J. (1996) 'How should national brands think about private-labels', *Sloan Management Review* 37: 89–102.

Howe, C. (1978) *China's Economy: A Basic Guide*, London: Elek Books Ltd.

Institute of Grocery Distribution (2001) *European Grocery Retailing 2001*, Letchmore Heath: IGD.

Koretz, G. (2002) 'Wal-Mart vs. inflation', *Business Week* May 13: 14.

Levy M. and Weitz, B. A. (1998) *Retailing Management*, Boston, MA: Irwin/McGraw-Hill.

Li, C. (1998) *China: The Consumer Revolution*. NY: John Wiley.

Lu, T. (2002) *Marketing in China*, Guangzhou: Guangzhou Press.

Miller, R. (1994) 'Crossing many borders', *Discount Merchandise* 34: 48.

Nolan, P. (2001) *China and Global Business Revolution*, London: Palgrave.

North, D. (1990) *Institutions, Institutional Change and Economic Performance*, New York: Cambridge University Press.

Ogbonna, E. and H. L. C. (2001) 'Power in the supply chain', *Customer Relationship Management* (March/April): 291–300.

Pellegrini, L. and Reddy, S. K. (1989) *Retail and Marketing Channels: Economic and Marketing Perspectives on Producer C Distributor Relationships*, London: Routledge.

Porter, M. (1985) *Competitive Advantage*, New York: Free Press.

Quelch, J. and Hardings, D. (1996) 'Brands versus private labels: Fitting to win', *Harvard Business Review* 74: 99–111.

Richardson, P. S., Jain, A. K. and Dick, A. (1996), Household store brand processes: A framework', *Journal of Retailing* 72: 159–85.

Rogers, D. (1991) 'An overview of American retail trends', *International Journal of Retail and Distribution Management* 19: 3–12.

Sethuraman, R. (1992) *Understanding Cross-Category Products*, Chicago MA: Marketing Science Institute.

Schultz, D. E., Tannenbaum, S. I. and Lauterborn, R. F. (1993) *Integrated Marketing Communications*, NTC Business Books.

Sternquist, B. (1997) 'International expansion of US retailers', *International Journal of Retail and Distribution Management* 25: 263.

Stiglitz, J. and Driffill, J. (2000) *Economics*, New York: W.W. Norton and Company Inc.

Stonehouse, G., Hamil, J., Campbell, D. and Purdie, T. (2000) *Global and Transnational Business: Strategy and Management*, New York: John Wiley and Sons.

Swinyard, W. R. (1997) 'Retailing trends in the USA', *International Journal of Retail and Distribution Management* 25: 244–50.

Treadgold, A. (1990) 'The developing internationalisation of retailing', *International Journal of Retail and Distribution Management* 18: 4–11.

Uhr, J. B. W. and Vering, O. (2001) *Retail Information Systems Based on SAO Products*. Berlin: SpringerVerlag.

Index

Printed in the United Kingdom
by Lightning Source UK Ltd.
124538UK00002B/67-78/A